The PhD Factory

The PhD Factory

Training and Employment of Science and Engineering Doctorates in the United States

Charles A. Goldman
RAND

William F. Massy
Stanford University

ANKER PUBLISHING COMPANY, INC.
Bolton, Massachusetts

The PhD Factory

Training and Employment of Science and Engineering Doctorates in the United States

ISBN 1-882982-36-3

Composition by Sheridan Books, Inc.
Cover design by Delaney Design

Anker Publishing Company, Inc.
176 Ballville Road
P. O. Box 249
Bolton, MA 01740-0249 USA
www.ankerpub.com

Table of Contents

About the Authors

Charles A. Goldman

Dr. Goldman is an economist at RAND, specializing in the economics of higher education and international trade. He conducts research on graduate education, the costs and financing of research, institutional strategy, and student financial aid. He collaborated with Dominic J. Brewer and Susan M. Gates to produce a new look at competition and strategy in U.S. higher education, *In Pursuit of Prestige* published by Transaction Press. He has recently published an examination of the federal system of funding for research, *Paying for University Research Facilities and Administration* (with T. Williams) through RAND. He holds a PhD in Economic Analysis and Policy from the Stanford University Graduate School of Business, and an SB in Computer Science and Engineering from the Massachusetts Institute of Technology.

William F. Massy

Dr. Massy is president of The Jackson Hole Higher Education Group, Inc. and a Stanford University professor. He has been active as a professor, consultant, and university administrator for more than 30 years. Dr. Massy earned tenure as a professor in the Graduate School of Business where he also served as director of the doctoral program and associate dean. Dr. Massy moved to Stanford's central administration as vice provost for research and later as acting provost. During his 13-year tenure as Stanford's vice president for business and finance and chief financial officer, he developed and pioneered financial management and planning tools that have become standard in the field.

In 1987, Dr. Massy founded the Stanford Institute for Higher Education Research in the School of Education and accepted an appointment as professor of higher education. His recent work has been on academic quality assurance and improvement, faculty roles and responsibilities, cost analysis, processes for resource allocation, and mathematical modeling of higher education institutions-including a full-scale computer simulation of university behavior under the title, *Virtual U*. Dr. Massy directs the project on educational quality and productivity at the National Center for Postsecondary Improvement, funded by the US Department of Education. He also serves on the University Grants Committee for the Hong Kong government. Dr. Massy holds a PhD in economics and MS in management from the Massachusetts Institute of Technology, and a BS from Yale University.

Preface

This book is based on a research project principally supported by the Alfred P. Sloan Foundation. RAND Education and the US Department of Education's Office of Education Research and Improvement (OERI) provided additional support. We are indebted to Michael Teitelbaum of the Sloan Foundation for his dedicated oversight and challenging comments throughout the project. Ralph Gomory and Jesse Ausubel at the Sloan Foundation assisted us with their insights at several stages.

We are grateful to Marc Chun and Beryle Hsiao for their expert research assistance and to Janice Pang for editing and producing the original technical report, from which much of this book is drawn. Roberta Callaway provided valuable administrative support throughout the research effort.

We thank the very capable and cooperative staff at the National Science Foundation Office of Science Resources Studies, who made data and analyses available to us. Without the work of that office, studies of this kind could not be done. In particular, the creation of the CASPAR CD-ROM data system greatly reduced the costs and difficulty associated with tabulating large data series for our modeling. We also acknowledge the assistance of the National Academy of Sciences Survey of Earned Doctorates project team for preparing the custom aggregations of time to degree data used in Chapter 10.

Throughout this project, we have benefited from comments and criticisms offered by many knowledgeable people in science and higher education policy. We thank Dominic Brewer and Ron Ehrenberg who each prepared a careful critique of the original technical report. The improvements in this book are, in part, a result of their comments.

The preparation of this book was supported by the Sloan Foundation, the RAND Arroyo Center, and RAND Science and Technology. Donna Keyser and Judy Lewis at RAND contributed substantially to planning and rewriting chapters for this book, making it accessible to a wider audience than the original technical report. Judy Rohloff worked on several aspects of the final manuscript, diligently formatting the many tables to make them clear and easy to read.

In addition, we appreciate input of Alan Fechter, Paul Romer, Roger Benjamin, Fred Galloway, Steadman Upham, Charlotte Kuh, and others. The work presented here is enriched by the discussions of seminar participants at RAND, National Bureau of Economic Research, American Academy for the Advancement of Science, American Council on Education, American Association for Higher Education, American Association of Public Policy and Management, Western Economic Association, US Commission on Immigration Reform, and Cambridge Decision Dynamics Colloquium on Systems Thinking in Higher Education.

We would like to extend our personal thanks to our spouses, Sally Vaughn Massy and Gary Reisch, who had to put up with the late nights and weekends of computer modeling and interpretation.

Responsibility for the conclusions, as well as any errors and omissions, rests with us, the authors.

Charles A. Goldman

William F. Massy

The PhD Factory

Part I

Results and Conclusions

1. The National Investment in PhDs

Is it possible that the United States could be producing too many PhDs? Or that 21st century society, with all of its scientific and technological promise, could not benefit from increasing numbers of advanced technically trained scholars? Stories of unemployment for science PhDs are widely reported in the press. Those stories may not be isolated instances. They may be indicative of broad trends facing the nation's most highly trained science and engineering elite.

Based on new research exploring the training and employment of science and engineering PhDs in the United States, this book predicts that with constant future resources and with constant domestic employment outside academe, academic departments will produce an average 20% to 24% annual excess of new PhDs over employment opportunities. These projections stand in sharp contrast to previous publications calling for sharp *increases* in PhD production (such as Atkinson, 1990; Bowen & Schuster, 1986; and Bowen & Sosa, 1989).

In particular, our models show that expanding university research and education will make the imbalance worse over the long run. On the other hand, expanding demand outside the university sector would improve matters if the jobs in question were those for which the PhD is cost effective—not just for the individual whose training has been highly subsidized, but for the society as a whole.

But in many fields the amount of such needed expansion seems unrealistically large given that the increases would have to be maintained year after year.

In addition to their implications for doctoral training in science and engineering, our models address additional questions that have been vexing higher education researchers and modelers. How do demands for teaching and research impact the ways that academic departments allocate resources among faculty, graduate students, and postdoctoral fellows? How can we predict the length of time it takes to attain a PhD and the percentage of students who actually earn the degree? How do we track the progression of faculty through their careers? Our models for departmental decision making, student attainment of the PhD, and faculty career paths, respectively, attempt to answer these questions. The models, estimation methods, and empirical results advance the state of the art in each of the three areas.

This chapter begins by explaining why understanding PhD training is important for academics, policy makers, and the general public alike. It continues with a brief overview of model structure and results: Federal support for science and engineering training will lead to the overproduction of PhDs unless significant changes are made to the operations of academic departments. The chapter concludes with a discussion of policy responses at the federal, state, and institutional levels, each considered in light of the results and intuition from the modeling.

Why should we care about PhD training and employment?

Many economists argue that society's investment in scientific and engineering research is an important positive determinant of its long-term growth. Thus, the US government subsidizes the graduate training of scientists and engineers in an effort to spur research for the public good. But if those highly trained graduates cannot find employment in the research and teaching careers for which they have been trained, this public investment would represent a waste of limited societal resources.

The problems with US graduate education and PhD labor markets also have important implications for related social policies, for example US immigration policy. Academic departments are turning more and more to overseas students to meet their graduate student needs. The federal government subsidizes the education of these students, as it does all science and engineering graduate students, despite the fact that current immigration policy does not allow these students to remain in the United States after graduation. Therefore, viewed from a systems perspective, policies that support training of PhDs, including large numbers of foreign PhDs, not only have little to do with projected 21st century demands for such scholars in the United States but conflict with other important social policies.

We do not argue that the "overproduction" of science and engineering PhDs confers no private or public benefits. Such talented and highly trained people will almost always find jobs, and usually good ones at that. The question is whether the jobs in question require four or more years of very expensive postbaccalaureate research training. For example, could a talented master's degree holder be trained on the job at far less cost? If the answer is "yes," the incremental cost of PhD training has been wasted.

One cannot gauge the PhD employment gap by observing whether doctorate-holders are unemployed, as some studies have tried to do (COSEPUP, 1995). Nor can one use the growth rate of PhD employment outside the academic sector. PhD holders can displace people holding lesser degrees in the same field, other things being equal. Because nonacademic employment will take up the slack left by the academic sector, predictions rooted in gross statistics on the growth of nonacademic employment represent self-fulfilling prophesies rather than evidence useful for policy determination. To use the nonacademic employment statistics appropriately, one would need detailed information about the jobs being taken by PhDs. Gaining such information presents difficult challenges, which is why our models of PhD production and university demand for PhDs add value to the policy debate.

None of this would raise policy questions if markets were perfect and public subsidies were not an issue. In perfect markets without subsidies, the private choices of individuals and employers—

including the decision about whether or not to invest in the PhD—would tend to maximize economic efficiency. But the PhD marketplace is far from perfect and private choices are by no means the only forces driving the system. We have become convinced that government policies, including but not limited to direct subsidies for doctoral training, have produced a system which in equilibrium tends to overproduce PhDs. We shall analyze the system presently, but first we examine how the federal government supports doctoral training.

Federal support for science and engineering PhD training

In American society, we do not presume to manage the career decisions of individuals. Yet in public service fields such as law and medicine, public associations and government agencies have adopted controls on the production of trained practitioners. Similarly, in direct and indirect ways, government supports the training of PhD-level scientists and engineers in the United States. This investment by society in research and development is claimed by some economists (notably Romer, 1990) to be the key to our nation's long-term growth.

According to conventional public goods theory, because funders of private research cannot capture all of the benefits that society will derive from their investments, they do not have the same large incentives to invest that society as a whole does. Therefore, society desires to increase the incentives of private firms to engage in research because these resources will be recovered through increased productivity in the rest of the economy. Society has several policy options to enhance private investors' incentives: patent protection, subsidy of public-interest research, subsidy of private research, and subsidy of key inputs to the research process.

Patent protection attempts to safeguard the important benefits of an invention for its inventor. But because patent protection is limited in duration, the benefits of new inventions eventually become available to society at large without cost. In the US pharmaceutical industry, for example, when a new drug is developed, its inventor can obtain patent protection for a period of 17 years. During that time, the inventor can market the drug exclusively, or can license the production of the drug. During this period of time, the marketer enjoys a monopoly in the drug, and can command a relatively high price. After the patent expires, others can produce the identical drug and market it also. This competition usually reduces the market price of the drug significantly. But the opportunity to command a monopoly price for a period of time provides an economic incentive to the inventor. These incentives are believed to spur innovation.

Another option for society is to subsidize research in the public interest. Although public interest research is often tied to the conduct of a government function, such as national defense or exploration of space, public-interest science research can spur the creation of new industries and new products. As part of the US space program, for example, many private firms were given government contracts to develop new materials and processes. In the 1960s, these contracts supported advances in a wide array of spin-off technologies.

The US government also subsidizes research in the public interest that is not primarily intended to advance government purposes. The National Institutes of Health, a federally funded agency, provides funding to universities and laboratories for a vast range of health-related research. Indeed university-conducted research on biology has formed the basis of the burgeoning biotechnology industry. Federally sponsored research is searching for treatments for human diseases and sequencing the human genome to improve fundamental understanding of how basic life processes work.

Privately conducted research has served a similar function in generating spillovers that can improve the products and services of other firms. While the transistor was developed at Bell Labs, the diffusion of this technology into the rest of the economy has expanded opportunities in many industries. The chain of development that started with the transistor at Bell Labs led to the integrated circuit at Fairchild and the microprocessor at Intel. Today, a huge spectrum of products is faster, smaller, and cheaper because of these inventions.

Unlike many other national governments that have directly

subsidized private research and development, the US government has so far resisted such direct subsidization of private research, although indirect subsidies for research and development have been proposed as part of the federal tax code. Subsidizing the key ingredients of the research process is another option for society. For example, the US higher education system subsidizes PhD training for scientists and engineers. Because these highly trained, research-oriented, personnel are crucial to the research process in any organization—public or private—this system reduces the costs of conducting research.

The cost reduction occurs because PhDs do not need to seek compensation in the labor market for the full cost of their training. Since doctoral students personally incur only part of the costs of the training, they are more willing to accept lower wages than, for example, MBA degree holders are. In fact, considering all fields of study, PhDs earn no more than graduates with other degrees do. Put another way, if PhD students had to pay the full cost of their training, few would enter science graduate schools at the wage levels prevailing today. The huge financial investment needed to pay the full cost of PhD training would not be matched by the future rewards.

This situation is tenable for two reasons. One explanation is that students get intrinsic satisfaction from earning the PhD, even if the labor market rewards are small. It is conventionally thought that someone earning a PhD in literature is pursuing that course of study because of a love of the subject, rather than expectation of financial compensation for the years of study. To some extent the same motivations apply in science, even though the market rewards are, on average, greater than in literature.

A second explanation is that PhD holders are compensated by means other than money. One form of compensation stems from attractive working conditions. PhDs often enjoy a high level of discretion in arranging their workdays. Professors, for example, have wide authority to control how, when, and even where they perform their work. This level of control can induce talented people to work for lower wages than they might otherwise demand in more structured careers.

The nonmonetary explanation notwithstanding, it is hard to

imagine PhD production on anything like the present scale without the public subsidies that greatly reduce the cost of earning a PhD across all fields. Thanks to a variety of funding mechanisms, including fellowships, teaching assistantships, and research assistantships, it is common for a graduate student to attend school without paying tuition. These fellowships and assistantships generally pay a stipend to the student for living expenses. In 1995, science and engineering graduate students received stipends of $1,000–$1,400 per month. These students often incur very few direct costs to attend graduate school, although they do have to give up the opportunity to work at a conventional full-time job for the years they spend in school.

The potential for under-employment of PhDs

US government subsidization of graduate education and university research has paved the way for a fundamental imbalance within the educational system, resulting in the very real possibility for under-employment of PhDs. As our model makes clear, the US system of graduate education makes the production of PhDs more a byproduct of federally sponsored research than a response to any real or even perceived need by society for such talent. This is because the academic system in US higher education permits graduate schools to finance and produce science and technology PhDs based on their own internal research and training needs. The need for teaching assistants to leverage faculty time in undergraduate courses provides a further impetus. In order to remain in balance year after year, the system depends on ever-growing resources. If resources fail to grow, the system overproduces PhDs. But even if the federal government does increase support for research and undergraduate education expands in conventional ways , the increased production of doctorates that results will ultimately worsen the PhD employment gap once resources level off again.

Innovation will not result from the production of highly trained scientists and engineers if those people cannot find employment in settings where they can educate and/or conduct research. In most fields, these opportunities have traditionally been in the academic sector—that is, within the very same system that makes decisions about admissions to graduate programs and the expenditure of government, corporate and institutional funds. But as our examination of the academic decision making process will make clear, the present system of US higher education is so geared toward turning out PhDs and so little concerned with how they will find the resources to make advanced contributions to society that overproduction is a likely result.

The potential for overproduction of PhDs is directly related to the opportunities that these highly trained people will find to fully employ their technical skills. We do not mean to imply that our nation or our world will have too many of these skilled people in an absolute sense, only that the systematic underemployment of PhDs in science and technology may represent a waste of society's limited resources.

While no one can accurately predict the future of the academic employment market, strong growth seems unlikely given past and future trends in the environment. Many states are shifting funding away from higher education toward other government priorities, such as corrections. And with the end of the Cold War, the federal government has already reduced defense-related research and development spending. Civilian R&D may be a further casualty of long-term federal balanced budgets.

Within academia, three trends suggest that there will be slow growth in openings for faculty positions. First, increased use of information technology and distance learning implies that faculty positions are likely to increase at a slower rate than enrollments. Universities and colleges are likely to be hiring information and media specialists to leverage the time of professors in producing courseware. Therefore, the number of openings for traditional faculty members is unlikely to grow strongly, even though demand for education increases.

The second trend that implies slower growth in faculty openings is that states experiencing the greatest growth in demand for education have extremely limited resources to allocate to expanding faculties. In states such as California, the state budget is increasingly allocated toward corrections, health, and welfare. Funding for K-12 education is constitutionally protected, meaning that higher education's funding is squeezed. Estimates developed at RAND project that higher education's share of the California State budget may drop by 50% in a decade (Carroll, 1995).

Third, just as PhD programs proliferated throughout the US academic system over the past few decades, the future may bring an increasing number of good quality PhD programs overseas. Asian countries in particular are contemplating enhancing local PhD training. The graduates of these programs will compete with US-trained PhDs, especially for overseas jobs. This trend may mean that fewer US-trained PhDs will find jobs overseas, further increasing pressure on the domestic US labor market for PhDs.

Growth of the domestic nonacademic employment market for science and engineering PhDs (mainly industry and government) represents the remaining variable in the equation. We have already argued that the growth should be in jobs where the PhD is cost effective. No presently available data, or to our knowledge any data planned for collection, can be used to predict the growth of such jobs. Therefore, the question of whether the growth of cost-effective nonacademic jobs will be sufficient to close the employment gaps we identify must be a matter of judgment. Our models process the available data in ways that sharpen the issues and thus open the way for coherent judgment. We believe that the answer is clearly "no" for a number of important fields, and that the case is problematic in a significant number of others.

Policy responses

Although the US system of higher education appears to be producing more doctorates than the market will bear, the solution does not lie in funding expansion—though more money for academic science and engineering maybe desirable on other grounds. Unless society is prepared to increase real funding year after year in order to keep the system in balance, which seems highly improbable, other mitigation for doctorate underemployment must be sought. A variety of policies both outside and inside the academic system present some opportunity for making that system more responsive, in appropriate ways.

Changes in government policy

Federal policy plays a fundamental role in shaping the academic system. Funding policies from the federal government, as well as immigration policy, interact with departmental decision-making to produce the conditions for underemployment of PhDs.

Federal funding policies, for example, give principal investigators broad discretion in employing personnel to assist on their research grants. Professors employ graduate research assistants, producing PhDs as a byproduct of their grant and research cycle. Limiting the funding of graduate students as research assistants and instead providing fellowships to appropriate numbers of promising students might allow the academic system to regulate itself more effectively.

Currently, US immigration policy makes entry to the US labor force difficult for foreigners. However, in the case of foreign students wishing to study in the United States, all that is required is their declaration that they do not desire to remain in the United States after graduation. As a result, the federal government subsidizes the education of these students, as it does all science and engineering graduate students, even though immigration policy does not allow these students—with the exception of exceptional scientists—to remain in the United States after graduation. Therefore, when viewed from a system perspective, government policies that support training of PhDs, including large numbers of foreign PhDs, are in conflict with immigration policy.

While fundamental changes in federal policies may be a long way off, some state governments are now attempting to direct how resources are used in their publicly supported institutions. In an effort to demand more accountability from public higher education, there are calls for universities to emphasize undergraduate instruction, caps on foreign enrollments, and resistance to subsidizing research infrastructure. In Florida, for example, state legislators have instituted required weekly teaching contact hours for faculty and increased restrictions on universities' use of state-supported faculty time for research. Such efforts seek to influence the mix of academic outputs paid for by public funds.

These rules may be considered extreme today, but will become much more prevalent over time. Indeed, as the imbalance between production and employment of PhDs grows, external pressures to change the system of graduate education will also increase. The academic system can wait until external mandates force change upon it, or it can take steps to change on its own.

Changes within the academic system

The first and simplest change to the academic system is simply to make the realities of science careers more apparent to prospective students. Understandably many prospective graduate students look at the careers of today's full professors and anticipate that they will experience a similar career. Even when the academic system was growing rapidly, less than a quarter of PhD graduates found positions in research university departments. Today, in most fields, even fewer will find such jobs.

A closely related recommendation is to modify graduate science and engineering programs to make the training more appropriate for a wider variety of careers. In science, the master's degree presently signifies failure rather than accomplishment, a consolation prize for students that do not progress to the PhD. As the academic system comes to a period more like steady state instead of steady growth, it is appropriate to consider advanced programs of education that prepare students to pursue careers not only in academic research but also in industrial engineering and science. And for jobs within academe, PhDs should be trained to teach undergraduates as well as to do research and constantly replicate themselves. Graduate departments have a responsibility to make the PhD relevant to the careers to which students will have access, rather than the relatively slim chance of a tenured faculty position directing a research lab. Moreover, training for a broader variety of jobs—not all of which involve replication—will reduce the pressure to overproduce.

Academia must also consider more sweeping changes in the way work is performed. It seems to us that the ultimate answer must lie in academic restructuring, in a change in the mix of human resources in academic production generally. More specifically, instead of relying on graduate students for such a large part of teaching and research, faculty must devise alternative ways of performing these functions, either themselves or by using other professionals. Technology offers the possibility of shifting work from graduate students to professional employees or technicians, but this will require a conscious effort by departments and university budgeting authorities. Such a restructuring will separate the demand for graduate students from the production of those students, as is fundamentally necessary and appropriate. Such changes in the internal workings of academic departments can help rectify the alleged over-use of doctoral students in undergraduate instruction.

Organization of this book

This book is organized into two parts. The first part (through Chapter 7) presents the development of the model and its results with a minimum of technical apparatus. The second part (starting at Chapter 8) gives full details of the modeling, data, and assumptions, and reports more comprehensive results.

Specifically, Chapter 2 presents a more detailed description of production structure and decision-making within academic departments, illustrated by quotations from a faculty-interview project. Chapter 3 explores further the intuition behind our PhD labor market model, followed by a summary of its structure and results. Chapters 4, 5, and 6 present a summary of the three submodels—departmental decision making, student attainment of the PhD, and faculty career paths—that feed into the overall labor market simulation. Chapter 7 draws together the important insights from this first half to shed light on the whole academic system and the wider social systems that affect and are affected by it. The remaining Chapters 8, 9, 10, and 11 cover in full detail the technical components of the simulation across the overall model and the three submodel components respectively. Chapter 12 highlights opportunities for future work. An appendix reports how each individual department at some 200 research universities was treated in the models.

References

Atkinson, R. C. (1990, April 27). Supply and demand for scientists and engineers: A national crisis in the making, *Science, 248,* 425–432.

Bowen, H. R., & Schuster, J. H. (1986). *American professors: A national resource imperiled.* New York, NY: Oxford University Press.

Bowen, W. G., & Sosa, J. A. (1989). *Prospects for faculty in the arts and sciences.* Princeton, NJ: Princeton University Press.

Carroll, S. (1995). *Projecting California's fiscal future*, MR-570-LE, Santa Monica, CA: RAND.

Committee on Science, Engineering, and Public Policy (COSEPUP). (1995). *Reshaping the graduate education of scientists and engineers.* Washington, DC: National Academy Press.

Romer, P. M. (1990, supplement). Endogenous technical change. *Journal of Political Economy, 96.*

2. The System of Training and Employing PhDs

The structure and organization of the academic department is critical to understanding how students are trained in science and engineering. Departmental structure is also the most likely locus of potential policy impacts. This chapter draws on a rich structured interview database with over 300 faculty members across the United States to paint a picture of production structure and decision-making in academic departments. The interview data help to build the case for the central hypothesis of the modeling: that doctoral students are recruited into departments to serve teaching and research needs and that the production of PhDs is a byproduct of faculty activities. The data further support the hypothesis that decisions about PhD production are not sensitive to external labor market conditions.

Although the interviews for the project was much more comprehensive, the quotations used here are from faculty members and department chairs in science departments that grant the PhD degree. The project did not include engineering departments, though we believe that much of the same dynamics operate in engineering PhD programs as in science PhD programs.[1]

For the purposes of this book, the department is considered to be essentially synonymous with its faculty, since the governance of a department is usually by consensus of the faculty members. However, since an individual department may depart somewhat from the general pattern of its college or university, the modeling of this book treats each discipline at each institution separately rather than relying solely on the classification of the whole institution. As a general rule, liberal arts colleges have departments that serve undergraduate students as well. Most departments in masters universities serve only undergraduate students, but some departments may serve graduate and professional students. Doctoral-granting universities and research universities have departments that educate undergraduate and masters students and also train PhDs.

Distinguishing features of graduate education

There are important distinctions in training and financing between undergraduate and graduate education. Undergraduates study a relatively broad curriculum consisting of subjects in a major field of study as well as general courses in a variety of arts and science disciplines. Full-time undergraduates are expected to spend all of their time either in the classroom or in preparation for related classroom assignments. Undergraduate students and their families typically pay for their educations or receive financial aid based on family financial need or the student's academic, extracurricular, or athletic prowess. Such financial aid consists of grants (scholarships), loans, and sometimes part-time work opportunities (work-study). While tuition discounting at the undergraduate level has become pervasive, the fact remains that most undergraduates pay some tuition and substantial numbers pay full tuition.

The education of graduate students, especially those in doctoral programs, is very different. Graduate students specialize in a single subject, taking increasingly specific courses related to current research topics. Doctoral candidates often spend half or more of their graduate school enrollment not taking courses at all, but conducting research under the supervision of a faculty dissertation advisor.

The differences between undergraduate and graduate education are financial as well as structural. Doctoral candidates in the sciences usually pay little or no tuition, and they receive a stipend for living expenses. These tuition waivers and stipends come in three major forms: fellowships, teaching assistantships, and research assistantships. Fellowships closely parallel undergraduate scholarships, generally providing for a waiver of tuition and a stipend payable monthly, quarterly, or annually. Teaching assistantships are arrangements whereby a graduate student, known as a teaching assistant or TA, teaches courses or portions of courses. In exchange for the TA's services, the department waives the tuition charge and provides a monthly stipend for the student's living expenses. A research assistant (RA) has a similar arrangement, except that the RA assists a faculty member in performing research in exchange for the tuition waiver and stipend. These forms of support for graduate students are awarded on the basis of academic merit. The student's or family's financial resources are not a factor in the decision to award a fellowship, TA, or RA.

For many students, entering a doctoral program, especially in science and engineering, amounts to a decision to be employed as a TA or RA while getting an advanced education. In the 1990s, stipends ranged between $1,000 and $1,400 per month. Although subject to state and federal income tax, stipends are generally exempt from other employment taxes such as Social Security, Medicare, and state disability insurance. The exemption from employment taxes is equivalent to $120–$150 a month in additional take-home pay. Graduate students are also allowed to defer student loans while in graduate school. These financial arrangements allow a student to pursue doctoral-level education in science or engineering while sustaining a modest lifestyle without any out-of-pocket cost for tuition, rent, food, or transportation.

Determining annual graduate student enrollments

A fundamental question for the functioning of the system of training and employment is how departments determine the number of new graduate students to admit each year. This determination and other elements of organization and decision-making in the academic department are illustrated here with quotations from the faculty interview project.

When asked how the number of graduate students is determined in a department, faculty cited three factors most commonly: the number of faculty who can advise students; the number of teaching assistants needed for undergraduate courses; and the amount of research money available to fund assistantships. Overall, the interview results indicate that departmental needs for research and teaching assistants drive the intake of graduate students and the resulting production rate of doctorates, and that departmental doctoral-student intake is limited by financial constraints rather than output-market considerations.

Departmental needs drive intake of PhDs

Departments are concerned that graduate students need mentoring and advising from faculty, therefore the number of graduate students per faculty member should not be too large. At the same time, faculty enjoy working with graduate students, so the number of graduate students ought to be large enough to give each faculty member the satisfaction of being involved in the graduate education of several students. A professor of computer science stressed the importance of striking a balance in the ratio of graduate students to the faculty members who will advise them:

> The factor that comes into the admission of graduate students is basically the number of faculty. We have 75 active graduate faculty. We asked, 'What is the reasonable number of graduate students a faculty member can handle?' We came up with the number 6. So 6 × 75 is 450. (*Professor of computer science*)

In order to make this calculation, the department has to determine the number of "active graduate faculty," that is, the number of faculty who will supervise graduate student coursework, research, and teaching. Departments may have other faculty that specialize in undergraduate teaching and are thus excluded from this calculation.

Teaching has a heavy influence on most departments' perceptions of their need for graduate students. There is a strong link between the number of teaching assistants needed by a department and the number of students admitted. An associate professor of chemistry explained:

"... how many teaching slots we have [is] ... directly tied to how many students we can admit into our graduate program." (*Associate professor of chemistry*)

The nature of a department's teaching also may influence the intake of graduate students. This same professor described the close relationship between a department's perception of its teaching needs and the intake of new graduate students:

> ... it comes back to this TA problem. Some departments have these huge classes that need a lot of TAs in the first year, and so they accept a lot of people into the program so they can teach all these things and then weed people out after the first year, because then they'll get the next class in. (*Associate professor of chemistry*)

A department's research needs also play a critical role in determining the number of students admitted:

> There's a real symbiosis here ... We need our graduate students every bit as much as they need us because they do our research. It's their hands, and it's their minds ... The students bring a freshness of approach, a lack of already tattered thinking that's very valuable. (*Biochemistry department chair*)

Graduate students bring two capabilities to the department. They do the research work and they offer a freshness that is important to keep advancing the state of knowledge.

Budgetary considerations are also cited as interacting with the teaching and research needs.

> I think the main ... criterion of how many people to admit is budgetary. It hasn't always been that way. The other factors used to be what is the size of the department, and would there be enough graduate students, number one, to man the elementary courses, number two to supply research, and those two things—and the number of applicants—decided how many to admit. But there has always been a budgetary component also ... (*Professor of physics*)

The characterization of the budget, though, depends heavily on the funding for positions of TAs and RAs. The bottom line in most departments is:

> You can't admit more than you can identify funding for. (*Professor of physics*)

This definition of funding includes fellowships and TA and RA positions.

PhD admissions insensitive to external labor market conditions

Although our interview respondents—both department chairs and faculty—did appear concerned about the employment opportunities for their PhD graduates, this concern did not translate into substantive changes in admissions policy. For the most part, departments focus their energy and attention on the increased number of degrees awarded by lower-status institutions and the identification of new job opportunities outside academia (or insistence that they do exist), rather than on a concerted effort to reduce their own enrollment in any significant way.

Many faculty members expressed concern about the labor market for their graduates:

> "... I do worry really quite a bit about [doctoral] students—getting them all positions. It's a very tight job market, and there are very very few people who are getting positions. I know this feeling is definitely held by many many of my colleagues. They are certainly concerned about how well their graduate students will find jobs ..." (*Assistant professor of physics*)
>
> The squeeze is felt. It's being felt also because its hard for people who are getting their PhDs—there aren't jobs. Some people are taking longer, and that's putting a squeeze on the number of new graduate students that we can admit ... (*Professor of physics*)

Nevertheless, faculty were reluctant to decrease enrollment at their own departments, though some suggested decreases could be made in other departments or at other institutions:

There are some things that are true about how humanities departments are run that is deserving of some criticism, like over-production of PhDs in subjects that nobody is interested in but academics. Some may say that is also true of physics, so I have to be careful, even though I feel that physics research is more valuable. (*Professor of physics*)

. . . [I] recently learned that the number of PhDs that we're producing per year in physics [nationwide] has risen from roughly 900 when I came to [university X] to something like 1,400 now. That's absolutely crazy. I don't know of anyone at a major research university who is producing more PhDs than they were [then]. In fact, if anything, funding has gotten tighter and people have fewer students. The answer is in fact, these people are coming from places that used to not have graduate programs in physics at all. And then these people are being dumped out on the pavement, so to speak. (*Professor of physics*)

Departments do sometimes adjust their intake of graduate students, although often by an informal decision-making process.

This is between me and the faculty member who is in charge of admissions. He says to me, "How many do you want?" . . . We both look at each other and say, "I think we should go down a little bit." I say, "Fine." And we go down a couple, 10% or so . . . (*Professor of physics*)

Some faculty members acknowledged that a market exists for PhDs, but claimed that they were already taking it into account or were identifying new opportunities in the market for their graduates:

I think in general there do seem to be mechanisms that affect PhD production that work regardless of whether the department makes it its priority. Graduate students tend to apply for things they believe there are jobs in and even if faculty may want more graduate students to continue their research programs, I think there's a general perception that if there are no jobs in that field, they are not going to get qualified applicants. (*Associate professor of psychology*)

Now there is no central control on the number of PhDs. It's a market. But the real problem is manpower prediction is impossible, truly impossible . . . almost every manpower survey that has been done has been wrong. So I think until we get much more accurate predictive capability, more complex systems . . . I wouldn't want to make any kind of short term corrections. *(Professor of physics)*

We are overproducing if the only thing a PhD in physics can do is to go into academics. But that is not the only option . . . It is our responsibility to make sure that not only do we meet their expectations but that we shape their expectations in ways that are realistic and sensible. You take people and make clear to them what is realistic . . . You make clear where there are exciting problems outside of a university. *(Professor of physics)*

One faculty member sought to resist short-term market pressures:

The bottom line is, just ask yourself what common sense tells you about whether we need more technically trained people or fewer in the coming generations. And the answer is clearly we need more. *(Professor of physics)*

In short, faculty express concern about the labor market for PhDs and will do what they can to place their own student—but their concern seldom leads to adjustments in doctoral student intakes. Faculty tend to believe that more scientifically trained manpower is better than less, and that job opportunities will somehow materialize. In any case, the department's short-run requirements for inexpensive research and teaching labor, and the desire of faculty to replicate their own skills, is of stronger relevance to admissions decisions than the more abstract and distant concept of labor market balance.

Quality of PhD applicants is largely irrelevant to admissions

Some professors referred to the quality of the applicant pool as a factor in determining the number of new graduate students enrolled each year. The department needs TAs and RAs, but it is also concerned about the quality of the graduate students admitted. "It is always an uncomfortable mixture of quality and the number of TAs/RAs needed. It has to be." *(Professor of Physics)*

Surely departments should cut back admissions if qualified applicants cannot be found. In other words, the doctoral labor market might be regulated by the supply of prospective students even if faculty do not shoulder the responsibility for regulating it themselves. For example, some respondents indicated that their graduate program intake is indeed limited by the number of applications received as well as by available funding:

> We like to let in students at a certain level, so we don't admit them if they don't come to that level. If we get more applications, we let more in. We have a big enough faculty so we can handle that. The factor that determines a lot is the number of teaching assistantships we have, which is down a bit because the state is running out of money . . . (*Professor and former department chair of math*)
>
> There's many more openings in chemistry departments, at least as far as I can tell, than there are students applying and going to graduate school. So there is competition for students. (*Professor of chemistry*)
>
> . . . the problem with our graduate program is not enough good applicants, pure and simple. I think that's a general problem that's not limited to University X. There are many universities that are not happy with their graduate applications . . . (*Professor of biology*)

Our database is silent on whether shortfalls in applications are caused by student concern about future job prospects, lack of preparation, or intrinsic disinterest in academic and research careers. However, we doubt that applicant-based regulation provides much of an equilibrating force in any case. First, our interviews suggest that while precipitous quality declines may cause departments to limit admissions, slow declines tend to be accepted as long as minimum quality thresholds are maintained.

Uncertainty about an applicant's potential and motivation compounds the admissions problem and also produces attrition:

> Ideally what we like in a graduate student, a beginning graduate student, is somebody who has very good research experience already, who can come into the lab and be productive right away. Well, we tend to find that students have very little experience and really need to be trained from scratch. In many cases, they haven't even really decided whether or not they want to be chemists. And that's why a lot of them quit. That's a problem. (*Professor of chemistry*)
>
> We could cut down the program, have fewer students,

without any loss—assuming we get rid of the worst ones. But well, you know, you get a whole lot of grad students anyway, it's not so easy to tell who's going to be good and who isn't when you bring them in. (*Professor and former department chair of math*)

We suspect that most departments will take higher risks as applicant quality declines, to the point where even the threshold norms may erode over time. Faculty may rationalize their behavior as giving more students a chance. Faculty may also take comfort in the fact that the truly unsuitable students will quit the program prior to the point where faculty members make large personal investments in qualification exams and dissertation research—but, perhaps, after those students have been of some use in furthering the department's teaching and research goals.

Respondents in many fields perceive the quality of US applicants to be declining relative to that of their foreign counterparts. The substitution can benefit the US labor gap but, absent extensive language and cultural training, foreign TAs may undermine the quality of undergraduate education. Departments are admitting more foreign applicants because of their perceived quality advantage, especially in research, though they try to maintain a balance with US students:

> We have a problem, there are three ways of cutting the group. There are bright students and less bright students. There are American students and foreign students. There are theorists and experimentalists. Unfortunately, all three ways of cutting give you the same group: by and large you see this pool of very bright foreign, theoretical students and less able American experimentalists. This is the way it is. But one has to have a reasonable balance. So it means that at some point, admitting these students means lowering your standards. (*Professor of physics*)
>
> We have about 400 applications, between 80 and 100 offers and about 50 acceptances this year—maybe a few less. We select by quality, but try to maintain the ratio between US and foreign students ... applicants are overbalanced with European applicants, so we are more lenient with American students. Currently, we have about 45% American students. (*Professor of physics*)
>
> There is one problem though and that is if I only look at

the grades and qualifications, we would have only foreign students. Now of course one tries to have some balance between the American and foreign students. It is a tough issue because very often the American students you admit are not the best students, and that leads to animosities and frustrations among the graduate students. (*Associate professor of physics*)

Suboptimal behavior of the graduate education system

Rewards for faculty and their departments lie primarily in the conduct of research as opposed to teaching. At the most prestigious universities, junior faculty are evaluated almost solely on the basis of their published research record in tenure decisions. Therefore it is not surprising that faculty tend to equate academic productivity with research outcomes (Massy & Wilger, 1995). And because graduate students in science and engineering are viewed by faculty primarily as an aid to conducting research, a department's ability to become or remain competitive in the national research market is thought to be inextricably linked to the admission of graduate students. As a result, deans, department chairs, and other central academic authorities find it difficult to decrease the intake of graduate students. This difficulty is compounded by the fact that the real determinant of graduate admissions is funding. If an able professor brings in research grants or contracts, then she or he can support the enrollment for some additional graduate students. Since the professional rewards for the professor and the prestige of the department and the college are advanced by admitting graduate students to participate in funded research, there is no incentive to regulate admissions.

Substitution of foreign for domestic applicants aside, we believe that strong forces work against the likelihood that PhD production will be adequately regulated through application-based intake adjustments. First, strong departments tend to be insulated from precipitous declines in the quality of the graduate student applicant pool simply because of their competitiveness. Moreover, they believe that graduate student intake reductions should take place in lesser-quality departments, not where doctoral training is at its best. Weak departments probably see larger graduate applicant quality changes but they are reluctant to reduce or eliminate doctoral programs because of the adverse effect on research. While a few departments have taken self-denying ordinances and reduced PhD production in recent years, such behavior is much more the exception than the rule.

In summary, strong research departments might be able to reduce the number of doctoral students in their production mix—in effect, by substituting higher-cost for lower-cost labor—without damaging their competitiveness. However, such internal restructuring is viewed as undesirable because it would lower the overall quality of doctoral education. Weaker departments and institutions would find such a substitution difficult, since they enjoy less prestige and operate on smaller financial margins. Hence they try to maintain doctoral programs even when reduction or elimination makes sense from a system-wide perspective. As a result, the system of graduate education is locked into a pattern of suboptimal behavior, inhibiting adaptation to labor market imbalances and leading to the unwelcome prospect of overproduction of PhDs.

Endnotes

[1]The Faculty Roles and Responsibilities project was funded by the Consortium for Policy Research in Education (CPRE) under a grant from the US Department of Education's Office of Education Research and Improvement (OERI), with William F. Massy as principal investigator. The Alfred P. Sloan Foundation also supported a portion of the fieldwork in connection with the project reported here. For description of the results of the larger project, please see: Massy, W. F., Wilger, A., & Colbeck, C. (1994, January–February). Overcoming 'hollowed collegiality'. *Change*, 10–20; Massy, W. F., & Wilger, A. (1995, July–August). Faculty productivity. *Change*.

In all, we interviewed 344 faculty at 19 institutions, of which nine grant the doctorate. The interview protocol was designed to elicit conversations with department chairs and other faculty on the subjects of interest rather than responses to a series of preset questions. The conversations were recorded and later transcribed into a database of "episodes," where each episode describes the discourse on a particulate subject. The database contains some 5,000 episodes, of which 500 pertain to the admission, training, or employment of graduate students. Departments in eight different subject areas (not all in science and engineering) are represented in the 500 episodes.

References

Massy, W. F., & Wilger, A. (1995, July–August). Faculty productivity. *Change*.

3. The Employment Gap for PhD Scientists and Engineers

This chapter explores further the intuition behind our PhD labor market model and presents a nontechnical summary of its structure and results. To simplify the analysis, we present results for a single field—mathematics. The model projects production of PhDs and academic job opportunities for PhD holders and concludes that stable resources lead to an overabundance of PhDs in mathematics at the rate of 32% of each year's production. The projection assumes constant nonacademic (i.e., industry and government) employment, but we shall see that no plausible growth rate of such employment could balance supply and demand. This chapter concludes with a summary of the results for all 12 fields covered in the study, which show employment gaps ranging between 0% and 45%.

The three subsequent chapters contain a nontechnical summary of the three submodels (departmental decision making, student attainment of the PhD, and faculty career paths) that feed into the labor market model.

The intuition behind the model

A few years ago, one of us served on an Office of Technology Assessment advisory panel whose charge was to assess the health of US academic research. Not surprisingly, conversation at one of the meetings came to lament the "underemployment" of young scientists and engineers—the difficulties faced by new PhDs in establishing themselves in suitable posts and developing independent research careers. One distinguished scientist asserted the answer to be for Congress to appropriate sufficient additional research funds to provide America's doctorate-holders the chance to pursue effectively the careers for which they are trained. The argument was couched in terms of the national interest as well as on social value grounds, since, with underemployment, a portion of the public investment in doctoral training is wasted.

While there can be no doubt that expanding sponsored research budgets increases the demand for doctorates, it is also likely that the process for supplying PhDs plays a role in creating the employment gap. There is no reason to believe that the demand effects outweigh the supply effects once doctorate production has had a chance to adjust. Could it be that, while salutary at the beginning, increased research funding actually makes doctorate underemployment worse over the long run? And what about subdoctorate student enrollments? While increasing the need for faculty, they also may call forth new demand for doctoral students to help teach elementary courses—students who will, in due course, attain the PhD and enter the doctorate employment market.

Simple calculations suggest that the idea of a supply-demand imbalance in PhD employment is not far-fetched. For example, if a moderately active doctoral-producing professor's career spans 40 years and he or she graduates one PhD every four years, the gross "reproduction rate" is 10 to one. Some of the PhDs will find jobs in industry or government or will enter academic institutions that do not produce PhDs. Some, though, will take positions in PhD-producing departments and continue the cycle. If only 15% enter PhD-producing departments, the "net reproduction rate" would be 1.5 to 1.0—50% greater than the 1.0 needed to maintain a stable population.

Our central hypothesis is that the US academic system causes PhDs to over-reproduce when demand is stable and resources are steady. Our argument can be summed up in two statements. First, the "natural production rate" of doctorates tends to exceed the average steady-state uptake for new faculty and postdoctoral positions by US institutions by more than the plausible growth of net demand from industry, government, and the rest of the world. (Exceptions occur in fields, like computer science, which are enjoying an explosion of industrial demand.) Second, growth of the US university system will absorb the excess doctorate supply during the expansionary period, but overproduction will reassert itself once the expansionary period has passed. In other words, US institutions tend to overproduce doctorates, and the overproduction cannot be cured by a one-time expansion of the system. If this hypothesis is correct, calls for increased PhD production to alleviate an immediate shortage will only increase underemployment in future decades. Rather, universities must change their natural rate of PhD production: that is, the rate of dependency of PhD enrollments upon undergraduate and postgraduate enrollments, R&D funding, and other departmental costs and revenues.

An examination of recent data shows this hypothesis to be well supported. Table 3.1 presents data for seven science and engineering fields. Faculty members in these fields produce an average of between 0.23 and 0.48 new PhDs per year. (The average covers professors in all kinds of four-year colleges and universities.) Assuming that recent trends hold, about 36% of new science PhDs and 28% of new engineering PhDs go into academia. Of those going into academia, 40% find positions in PhD-granting departments. As the table makes clear, the career replacement rates exceed one in every field. This provides strong prima facie evidence that the system over-reproduces PhDs. Our work, which refines these simple calcu-

Table 3.1 Simple Calculations of PhD Replacement Rates

Field	PhDs per year per faculty	Fraction of PhDs to academia	Fraction of academics in PhD depts.	Annual replacement	Career replacement
Physics	0.27	0.36	0.40	0.039	1.56
Chemistry	0.52	0.36	0.40	0.075	3.00
Math	0.23	0.36	0.40	0.033	1.32
Chem. Eng	0.48	0.28	0.40	0.054	2.15
Civ. Eng.	0.32	0.28	0.40	0.036	1.43
Elec. Eng.	0.37	0.28	0.40	0.041	1.66
Mech. Eng.	0.31	0.28	0.40	0.035	1.39

lations by modeling the PhD production process and disaggregating institutional segments, leads to the same conclusion.

As described in Chapter 2, academic departments formulate their requirements for doctoral admissions based on the faculty's perceived need for research and teaching assistants and internal financial constraints, not from assessments of the external labor market for doctoral graduates. While science and engineering faculty do express concern about their graduates' employment prospects, these concerns do not dominate department decisions about doctoral-program size. The relation between doctoral applicant pools and labor-market conditions also seems to be loosely coupled—at least according to faculty perceptions. The faculty we interviewed were satisfied enough with applicant pool numbers and quality to continue with their doctoral enrollment plans. In short, the US system of doctoral production appears to operate in accordance with internally generated agendas of academic departments without much regard for external labor market conditions.

Under such circumstances, the existing system of graduate education can be balanced only when academic resources or demand from the rest of the world are growing steadily. Fluctuations in that growth, or level resources rather than growth, result in imbalances between the labor market supply and demand of PhDs. While there are some market mechanisms that may potentially improve the functioning of this system and hence result in a better balance of supply and demand, these mechanisms depend on the appropriate feedback of labor market conditions into decisions made by individual faculty, departments, and universities. Over the long term, labor market conditions probably do feed back into the academic decision making process, but as Chapter 2 reveals, powerful qualitative evidence suggests that the current system resists this sort of feedback. Faculty are mostly concerned about producing their own teaching and research and see graduate students mainly as low-cost inputs to that process.

The structure of the model

Our simulation estimation of the PhD employment gap depends on three submodels. The Departmental Choice Model predicts professorial hires, PhD admissions, and postdoctoral appointments as functions of the set of "driver variables" to be defined below. The Doctoral Attainment Model predicts completion rates and time to degree by student segment. The Faculty Transitions Model predicts departures from the labor force. We shall describe each model briefly here before discussion how they come together in our simulation of the employment gap. More detailed treatments can be found in later chapters.

The insights of Chapter 2 combined with extensive statistical data form the basis of the Departmental Choice Model. The building blocks of that model are shown in Figure 3.1.

Each department determines its teaching requirements on the basis of undergraduate and graduate student demands. Undergraduate student demands include those students majoring in the department, as well as general education demands from students majoring in other fields. For example, mathematics has a high general education demand because many students commonly take courses within this field. Majors rather than nonmajors, on the other hand, normally take chemical engineering courses. Graduate students in master's, professional, and PhD programs also generate demand for teaching.

Each department determines its requirements for research on the basis of available funding. Research sponsors include the federal, state, and local governments; private foundations; corporations; and individuals. All of these taken together add up to a de-

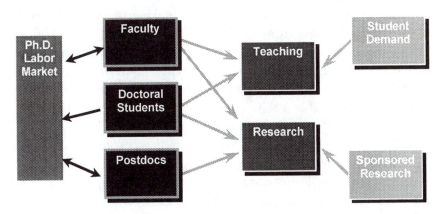

Figure 3.1   Overall Model Structure

partment's sponsored-research funding. The accounting system of universities means that sponsored-research funding must be applied to cover the costs of doing research. The university cannot receive these funds without documenting a corresponding equivalent expenditure. A portion of this expenditure is in the form of "indirect costs" which are the general costs of supporting research, including facilities and libraries, rather than "direct costs" of each project, which include individual faculty salaries, equipment, and graduate research assistants.

In order to meet its teaching and research needs, a department can use the labor of its faculty (tenure-line or adjunct), graduate assistants (TAs or RAs), or postdocs. Faculty and graduate students participate in both teaching and research, although any one individual may specialize in just one area. Figure 3.1 indicates that postdocs do not customarily participate in teaching; they are exclusively junior researchers.

An important departmental decision is whether to augment the existing pool of human resources with new hires or new admissions. Based on the current needs for teaching and research and the existing stock of faculty, graduate assistants, and postdocs, the department determines needs for additional faculty, additional graduate students, and additional postdocs. These demands for faculty and postdocs are then aggregated with all departments in that field (along with estimates of net demand from US industry and government and foreign employers) to estimate the total inflow and outflow to the PhD labor market.

Each year, graduate students either make progress or drop out of school. Using data from 25 years of PhD granting in the United States, the Doctoral Attainment Model estimates the progress and dropout rates for various fields and types of departments. This model allows us to simulate the number of graduate students remaining at the end of each year in each department and also the number of new PhDs who will emerge from the pipeline each year.

The Faculty Transitions Model predicts professorial career progress, midcareer departures, and retirements. The data here are much sketchier and our model is far less detailed and precise than the model for graduate students. Nonetheless, we can make some estimates of faculty turnover which allow us to simulate the number of faculty in each department at the end of each year. We also use alternative assumptions to minimize the chance that our model is misleading.

The simulation model calls the three submodels each year. Their inputs include updated demands for teaching and research and new estimates of the system's human resources. The Departmental Choice Model generates the new demands for faculty hiring and graduate student admissions. The Doctoral Attainment Model predicts the number of new PhDs and the Faculty Transitions Model predicts the number of departures from the labor force. These quantities, along with the nonacademic demands which are determined exogeneously, determine the estimated balance between demand and supply in the PhD labor market.

Because the level of student enrollments determines teaching demand and sponsored-research funding determines research demand, we refer to these two values as the "drivers" of the model. Formally, student enrollments consist of several values: general undergraduate enrollment, undergraduate majors in that department, graduate enrollments (not including PhD students) and PhD students. However, because PhD enrollments are determined endogenously, that is, *within* the model, we do not use PhD enrollments as a *driver*. Therefore the drivers (for each field) consist of four values: general undergraduate enrollment, undergraduate majors in that department, graduate enrollments (not including PhD students) and total sponsored research. These values are captured over time and also disaggregated by departments, but we will deal with those complexities in the later chapters. We have collected data for 12 separate academic fields—namely bioscience, chemical engineering, chemistry, civil engineering, computer science, economics, electrical engineering, geoscience, mathematics, mechanical engineering, physics, psychology—across virtually all doctoral and four-year nondoctoral granting colleges and universities in the United States. We model each of these 12 fields separately. There is no overall aggregation in the model; each field stands on its own. In this chapter we use mathematics to illustrate the calculations of the model. Although the mechanics of the model are similar for all fields, the results will vary because of different driver values and demands for teaching and research in each field.

Results for mathematics

Driver values

The simulation system depends on the values of the drivers: student enrollments (excluding PhD students) and amount of sponsored research. In the field of mathematics, these two drivers have behaved very differently. Sponsored research grew (in real terms) over the 1980s, then leveled off, as shown in Figure 3.2. Undergraduate degrees in mathematics have held constant or declined over the recent past, as shown in Figure 3.3. The basic simulation that we present first assumes that these drivers will remain constant (in real terms for research dollars) over the future, which is a plausible assumption given the recent history in the field. Although we do not show the data here, total undergraduate enrollment in all majors is an important driver for mathematics because so many students from all fields need mathematics courses.

The model assumes that institutions try to respond quickly to changes in the driver variables. For example, an upward shift in undergraduate enrollments or research funding generate new faculty hiring and doctoral admissions the following year. The response to downward shifts can take longer because of rigidities in faculty em-

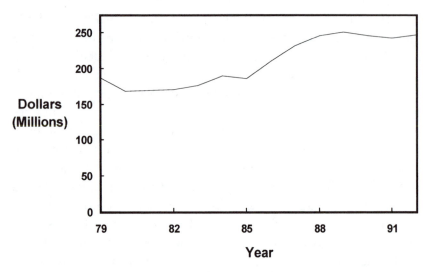

Figure 3.2   Sponsored Research in Mathematics

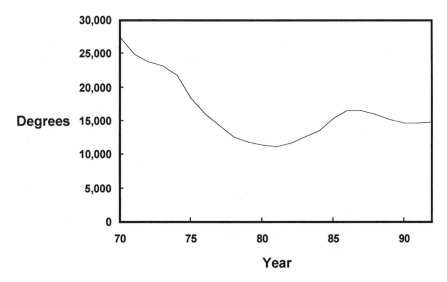

Figure 3.3   Bachelor's Degrees in Mathematics

ployment and the delays associated with doctoral training. The delay structure accounts for the simulation's asymmetric response patterns, but it has no effect on the base employment gap.

Base employment gap

Figure 3.4 illustrates our conception of the base employment gap, as it would appear when the system reaches equilibrium. Labor supply consists of new PhDs and faculty labor force re-entries. Demand consists of new faculty hires plus uptakes from "industry/other" and foreign entities. New faculty hires depend on the total demand for faculty, which depends on the driver, and net faculty attrition. The postdoctoral appointment flows are not shown, since they exactly balance each year with as many PhDs taking new postdocs as those completing postdocs and returning to the job market. Doctoral admission numbers determine PhD output in future years.

Because we cannot model industrial/other demand given current data, our calculation of the base employment gap assumes that such employment remains constant over time. Attrition is taken as 2.75% per year, which equates to an average career length of 36

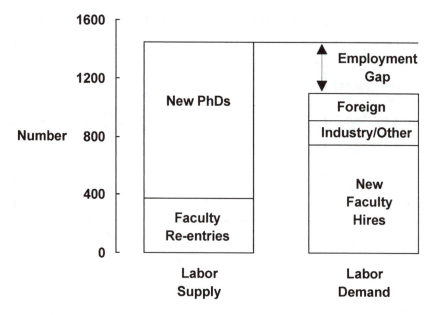

Figure 3.4   Employment Gap for Mathematics

years. Foreign demand is modeled by assuming that 50% of foreign doctorates become employed outside the United States.

If the level of student enrollments and amount of sponsored research remain stable, the new steady state of the system results in a significant employment gap for PhDs in mathematics. We estimate the gap to be between 32% and 47% of annual output. (The difference between the estimates is described in Chapter 8.) We also calculated the sustained annual growth of industrial/other employment ("breakeven" annual nonacademic employment growth) that would be needed to absorb the excess production on a permanent basis. That figure lies between 33% and 47%, which is clearly beyond the realm of possibility.

Sensitivity to driver changes

To explore the effects of increases in funding or enrollments on research and teaching demand, we conducted two additional simulations. These two simulations explore the range of behavior of the system in each field and offer insights as to how those behaviors would respond to other possible changes in the driver values. The two simulations treat first research funding and then student enrollments. Each simulation provides for a five-year expansion of demand in each field, simulated at a 2% per year real increase over the five-year period. The simulation stabilizes resources at the new higher level after the fifth year, thus representing a real and permanent increase of 10%. Atkinson (1990) argues that 4% annual real growth was the most realistic projection for the 1990s, "in order to maintain real [economic] growth and competitiveness." We chose to be more conservative. However, our results would have been qualitatively the same had we used Atkinson's number.

During the expansionary period, more employment opportunities are created as departments hire additional faculty. This drives the employment gap down in the short run. But departments also increase the size of their graduate programs, admitting more graduate students to teach or conduct research. So even though the system stabilizes with a higher level of resources, the employment gap is worsened. Once real resources stop growing, the creation of new faculty positions slows, which drive the employment gap upwards. Because of differences in the use of graduate student labor for teaching and research, each field responds differently to these resource expansions.

Figure 3.5 graphs the response of PhD employment in mathematics to assumed exogenous increases in sponsored research (*R&D*). Figure 3.6 shows how employment responds to increases in (nondoctoral) student enrollments for mathematics. Doctoral student enrollments are determined endogenously within the model. In each graph, the y-axis represents the change in employment, in terms of the number of PhDs employed. A positive value indicates improving employment prospects; a negative change means worsening-employment prospects. Each simulation holds the drivers at their base values for two years to establish a zero point, and then simulates the 5-year ramp of driver increases plus an additional 15 years to establish the new equilibrium.

As the figures make clear, increases in resources make the job market much looser in the short term. But, as the drivers level off, the employment gap returns, even worse than before. This is despite the fact that the system as a whole has 10% greater resources than it did before the expansion. In other words, this kind of ex-

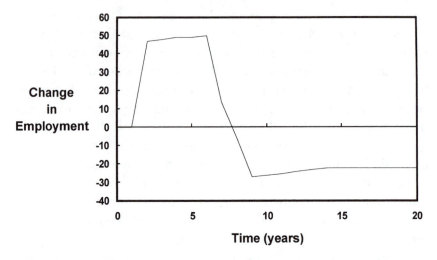

*Figure 3.5   Change in the Math Employment Gap with Increase in
Sponsored Research*

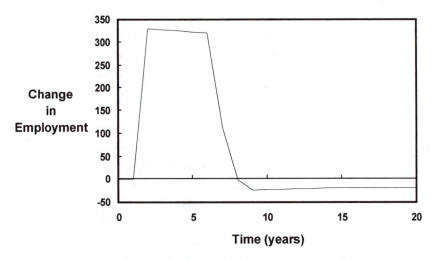

*Figure 3.6   Change in the Math Employment Gap with Increase
in Degrees*

pansion followed by a return to level resources (at a higher level)
worsens the employment gap.

Comparing the two graphs in Figures 3.5 and 3.6, it is easy to
see that the driver changes have markedly different size effects on
changes in PhD employment in mathematics, at least in the short-
term. Apparently, increases in student enrollments and the result-

ing increase in teaching demand are a more important determinant of PhD employment in mathematics than are increases in sponsored research. Indeed, the short-term response to the increase in student enrollments is about six times greater than that for the increase in sponsored research. The long-run responses are similar: a slight worsening of employment prospects for either simulation. In other fields the long-run responses may differ markedly as well, depending on the departmental decision-making structure.

Other factors such as PhD graduation rates have equal effects on the two simulations, though differences in these other factors are responsible for variations across fields. Chapter 8 contains a more technical exposition of the model and its results for all 12 fields.

Summary of results for all fields

Table 3.2 presents the estimated overproduction of PhDs for each of the 12 fields, as a percentage of the equilibrium number of PhDs produced annually. Similar to the aforementioned computations for mathematics, we present a high and a low estimate for each field. The figures are based on conditions extant in the early 1990s. They do not take into account any future employment increases. Although this is certainly not a valid assumption for every field at this point in time, we believe that it is cause for concern if the academic system systematically overproduces PhDs when supplied with stable resources. As Table 3.2 shows, 11 of the 12 cases we studied show this property.

The unemployment gap is largest for mechanical and electrical engineering at 44% and 41% respectively. Civil engineering, mathe-

Table 3.2   *Estimated Long-term Employment Gap*

Field	% gap		Field	% gap	
Mechanical Engineering	44	46	Geoscience	23	22
Electrical Engineering	41	41	Economics	23	27
Civil Engineering	33	29	Physics	9	12
Mathematics	32	46	Computer Science	4	5
Bioscience	28	41	Psychology	4	3
Chemical Engineering	26	29	Chemistry	(5)	(4)

matics, biosciences, and chemical engineering also show gaps in excess of 25%. The gaps for computer science, psychology, and chemistry are essentially nil. Since the results for different fields are derived from independent data sets, the case for labor imbalance in the science and engineering PhD system is a strong one.

Table 3.3 presents estimated growth rates of non-academic employment that would be needed to close the base employment gap for each field. The table shows that while the large figures we reported for mathematics were outliers, there are a substantial number of fields for which the breakeven growth rates seem implausibly large. For example, we doubt that cost-effective PhD nonacademic employment in mechanical engineering is growing to any significant extent, let alone at more than 8% per year. (Recall that the growth rate refers to jobs for which the PhD would be cost effective in a world without subsidies.) Electrical engineering is now riding a crest of popularity, but can the 50% increase in cost-effective PhD employment in ten years that is implied by the 4.8% annual rate be realistic? Even the figure for bioscience, another hot field, appears on the high side: will cost-effective PhD employment grow between 33% and 50% in a single decade? Nonacademic demand in chemical engineering, geoscience, and economics would have to grow by 16% to 33. The lower figure, associated with chemical engineering, might conceivably be possible if genetics-based products take off, but the other figures appear to be unrealistic. Consistent with the results in Table 3.2, the last four fields require essentially no growth to break even. (They show no employment gap either.) Indeed, there are likely to be shortages in computer science.

Table 3.3 Estimated "Breakeven" Nonacademic Employment Growth Rates

Field	% growth		Field	% growth	
Mechanical Engineering	8.3	8.6	Geoscience	2.0	1.9
Electrical Engineering	4.8	4.9	Economics	2.5	2.9
Civil Engineering	3.7	3.3	Physics	0.5	0.6
Mathematics	32.5	47.1	Computer Science	0.2	0.3
Bioscience	2.9	4.3	Psychology	0.2	0.1
Chemical Engineering	1.5	1.6	Chemistry	(0.2)	(0.2)

Table 3.4   Actual Growth Rates of Nonacademic Employment: 1993–97

	% growth			% growth	
Field	All	PhD	Field	All	PhD
Mechanical Engineering	2.9	4.7	Geoscience	1.0	1.6
Electrical Engineering	3.6	6.9	Economics	1.9	3.4
Civil Engineering	1.7	(0.4)	Physics	3.9	5.1
Mathematics	0.5	3.8	Computer Science	5.2	12.2
Bioscience	1.8	2.6	Psychology	1.3	2.6
Chemical Engineering	0.0	1.8	Chemistry	1.8	1.6

Additional insights about the breakeven growth rates can be obtained by looking at the actual growth rates of nonacademic employment for postbaccalaureate degree holders. Table 3.4 presents growth rates for the period from 1993 to 1997, obtained from NSF's Science and Engineering Statistical Data System (SESTAT).[1] The growth rates are for all postbaccalaureate degree holders and, for comparison, holders of the doctorate. The potential for substitution at the margin between masters and doctoral degree holders suggested that we look first at the sum of the two: that is, the "All" columns in the table. The actual growth rates are about half or less of the breakeven rates in six of the 12 fields—in six of eight if we exclude the fields with no significant base employment gap. This lends support to the argument that many of the breakeven rates are unrealistic.

The growth of PhD employment is larger than the overall rate for postbaccalaureates in all fields except chemistry, and there is it close. It is less than the breakeven growth rate in five of the eight fields with base employment gaps. (The positive difference is small in chemical engineering and fairly small in economics.) In any case, though, we put little credence in these statistics for nonacademic doctorate employment growth for the reasons discussed earlier. The growth of doctorate employment in the four fields without employment gaps, and to some extent in Electrical Engineering, do point to robust demand in those areas. The displacement of qualified masters-degree holders is not a problem when the base employment gap for PhDs is zero or negative.

Differences in employment gaps across fields can be linked to several factors in the model: each field's unique structure of departmental decision-making especially in terms of critical factors for admitting graduate students and hiring faculty, graduation rates for PhD students, and faculty career patterns. The next three chapters explain some of the results on field differences in each of these parts of the model. In addition, Chapter 8 reports in more detail the results of the simulations, showing sensitivity to increasing sponsored research or student enrollments for each field.

Endnotes

[1]National Science Foundation, Division of Science Research Studies [http://sestat.nsf.gov, Table C02]. To be consistent with our models, we calculated nonacademic employment as total employment minus employment in four-year colleges and universities. The figures are arithmetic averages of the changes between the surveys for 1993 v. 1995, and 1995 v. 1997. (Using averages reduces the effects of sampling fluctuations, although the differences between them and the 1993 v. 1997 growth rates are inconsequential.) SESTAT changed its sampling methods in 1991, so longer series are difficult to obtain.

References

Atkinson, R. C. (1990, April 27). Supply and demand for scientists and engineers: A national crisis in the making, *Science, 248,* 425–432.

4. Production of Teaching and Research in Departments

Building on the discussion of the academic department in Chapter 2, this chapter presents in an intuitive fashion our model of how departments allocate resources among faculty, graduate students, and postdoctoral fellows to meet teaching and research demands. The formal modeling is treated in detail in Chapter 9.

In Chapter 2, we heard from faculty that departments have certain perceived needs for teaching and research. Based on those needs, they determine the number of faculty, graduate students, and postdocs they would like to have in the department. The model formalizes this linkage between the demands for teaching and research and the human resources under the department's control.

Modeling the production and teaching of research

Figure 4.1 presents these departmental linkages in the form of a block diagram, departmental activities, the demands which drive those activities, and the human resources required to meet the demands. There are two types of demands on departments: a demand for teaching and a demand for research. Four types of student enrollment levels drive a department's calculation of its demand for teaching: undergraduate majors, undergraduate nonmajors, master's, and doctoral students. In estimating the combined demand of

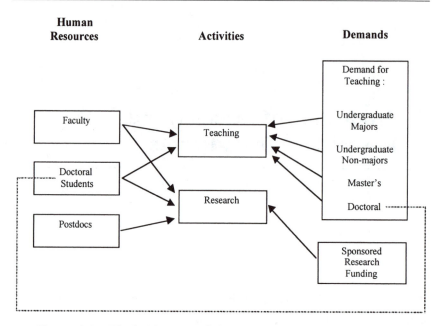

Figure 4.1   Block Diagram of the Departmental Decision Model

undergraduates who major in the department and undergraduate nonmajors, we use the total undergraduate enrollment at the department's whole institution. The model estimates the average teaching loads generated by majors and nonmajors. With regard to a department's enrollments at the master's and doctoral level, the model assumes that the department has control over setting the number of doctoral enrollments. For this reason, doctoral teaching demand is linked to doctoral enrollments in the diagram. Therefore, of the four sources of demand for teaching in a department, the model treats three as predetermined (undergraduate majors, nonmajors, and master's) and one as internally determined (doctoral). While a department can employ either faculty or doctoral students to meet demands for teaching from undergraduate majors, nonmajors, and master's, only faculty can serve the demand for teaching doctoral students. This is the one nuance in departmental operations not depicted in the block diagram of the model.

A department's demand for sponsored research is measured

by the total budget of sponsored research projects. The department can employ faculty, doctoral students, and postdocs to meet this demand. One of the strengths of our model is that it allows academic departments to vary in their patterns of human resource use—much as departments do in real life. Specifically, we classify all departments within each academic field into segments of similar types, namely more elite, typically research-intensive departments and less elite, less research-intensive departments. Institutions that do not grant the PhD are grouped into three large segments: public comprehensive universities, private comprehensive universities, and undergraduate colleges. These institutions follow the basic form of the model, but are restricted. Since they do not have doctoral students or postdocs, these universities must rely entirely on faculty to meet their teaching and research demands. Also because research is relatively unimportant in these segments, we model these segments as generating employment for faculty as a function of demands for teaching only.

How physics departments respond to changing demands

The model allows us to see how departments across different fields respond to increases or decreases in the research and teaching demands they face. In particular, because we segment departments within each field, we can gain important insights into how more elite and less elite departments differ in their patterns of hiring and admissions in response to demand changes. These findings, in turn, will shed light on how specific department types might alter their operations to produce an amount of PhDs more in line with available employment opportunities. We begin by considering how changes in the demand for research and teaching affect the human resource requirements for just a single field—physics. Afterward we summarize the results for all fields under study.

Figure 4.2 shows the estimated responses of physics departments to an increase in research funding of $1 million 1980 dollars. The graph separates the responses of the more elite and less elite groups of physics departments. Other departments fall in between

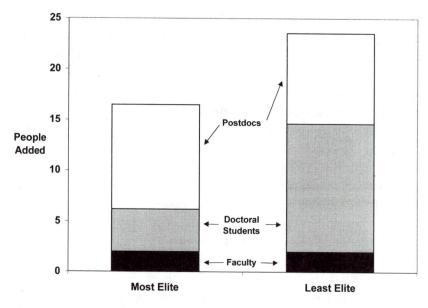

Figure 4.2 Hiring and Admissions Response to $1 Million Increase in Research Funding (Physics)

these two extremes. Departments without PhD programs are not included in these results.

As the graph indicates, a $1 million increase in funding produces about two new faculty positions in both more elite and less elite departments. However, the same funding increase produces a markedly different response in terms of doctoral admissions across department types. The more elite departments would admit about four new doctoral students, while the less elite would admit as many as 13 new doctoral students. There are two fundamental reasons why elite departments admit fewer doctoral students in response to the same increase in funding. First, more elite departments operate in more "expensive" environments. They have higher salaries, plus greater costs for research facilities and general overhead, so less money is available to employ doctoral students as research assistants. Second, more elite departments do not tend to increase faculty size and doctoral program size together, while less elite departments do. As described in more detail below, more elite, research-intensive departments with large faculties and more graduate students can hire fewer new faculty and admit fewer new graduate students relative to their less elite counterparts in order to meet increased demands for teaching.

The distinction in behavior between more and less elite departments does not apply to postdoc employment. The more elite departments hire about 10 new postdocs in response to the research funding windfall, and the less elite departments hire about nine. The cost factors described above apply to postdoc and doctoral student employment equally. A main reason the less elite departments actually hire fewer postdocs in response to increased funding is that many departments in this segment do not employ postdocs at all. Since our model extends the basic structure of each department's human resources from 1980 into the future, it generally assumes that departments which did not use postdocs in the past will not use them in response to increased resources in the future. Although the model allows deviations from the historical pattern, these deviations are unlikely to be large.

Now we turn to the effects of increased demand for teaching on physics departments' human resource decisionmaking. In accordance with the model illustrated in Figure 4.1, demand for teaching is comprised of undergraduate majors and nonmajors and master's components. Since doctoral teaching demands are internalized in the model, these are not included in our calculation of the net effects of changes in external demands.

Figure 4.3 reveals that for an increase in the size of the undergraduate student body that results in 1,000 more bachelor's degrees awarded each year in all subjects, physics departments in both the more and less elite segments would hire about three new faculty. The response in graduate student admissions, however, varies markedly by segment type. In the more elite departments, the increase in undergraduate student enrollment leads to admitting about 35 new doctoral students. In the less elite departments, the model estimates that about six new doctoral students are admitted. The explanation for this significant disparity appears to be quite simple: In less elite departments the primary teaching responsibilities fall on faculty, whereas in more elite departments doctoral students serve most of the general teaching demand generated by undergraduate nonmajor enrollment.

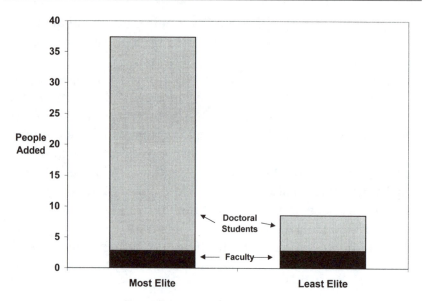

Figure 4.3   Hiring and Admissions Response to 1,000 Increase in General Undergraduates (Physics)

Interestingly, we find no distinction between more and less elite departments with regard to meeting the teaching demands of undergraduate physics majors. When considering an increase in major enrollment that would result in 10 new physics bachelor's degrees awarded each year, we find that departments, on average, would hire about two new faculty members and admit six new doctoral students. There was not enough variation in the data to distinguish differences by type of department, so we report all PhD-granting departments as behaving the same for this demand.

For master's enrollments, we follow the same modeling scheme as for undergraduate majors: We consider an increase in enrollments that would result in 10 more master's degrees in physics each year. Although Figure 4.1 accounts for doctoral students helping to meet teaching demands generated by master's students, in practice we found it impossible to run our model with this linkage intact while still separating the human resource decisions of more and less elite departments. Therefore, we decided to restrict our model so that only faculty could teach master's classes—a plausible though somewhat extreme assumption. As a result, we found that

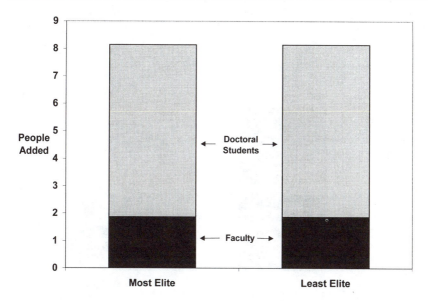

Figure 4.4   *Hiring and Admissions Response to 10 Increase in Undergraduate Majors (Physics)*

in the case of more elite departments, the increase in master's enrollments has the effect of increasing faculty hiring by about 1.3; in the less elite departments the effect is about 0.6. This is because less elite departments are likely to have larger classes and better established Master's programs in this field. In contrast, more elite departments tend to focus on research and the PhD rather than on teaching at the master's level. So to the extent they have master's students, it takes more elite departments comparatively more resources to teach them in smaller classes.

Summary of results for all fields

Our study uses the departmental decision model to examine the hiring and admissions decisions of typical graduate departments in 12 separate academic disciplines. These decisions are examined in response to the same changes in demand previously presented for physics departments alone: an increase of $1 million in research funding, an increase of 1,000 general undergraduate degrees, and

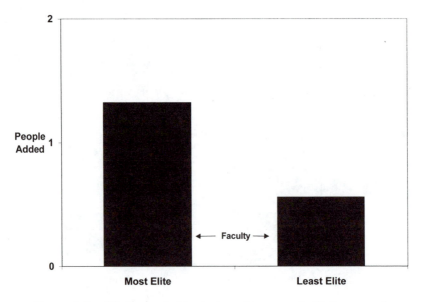

Figure 4.5   Hiring and Admissions Response to 10 Increase in Master's (Physics)

an increase of 10 undergraduate major degrees. We omit the master's degree information here because in some fields this source plays no role at all in changing the demand for teaching. The complete results of the model including differences between more and less elite segments across all 12 fields, are presented in Chapter 9. Here we show the average results for each of the 12 fields.

In general, we found a wide variation in hiring and admissions decisions across the 12 fields. As Figure 4.6 illustrates, in response to an infusion of $1 million in research funds, most fields would expand their graduate populations much more than their faculty. Indeed every field does some hiring of faculty and all but one admit more graduate students in this scenario. The impact of a research funding increase appears to be most significant in the field of mathematics, where the typical department would hire 22 new faculty, admit 55 doctoral students, and hire two postdocs in response to an influx of $1million in research funds. As it turns out, these hiring and admissions decisions would actually cost these departments even more than the million dollars received, forcing them to subsidize their human resource expansion through other activities.

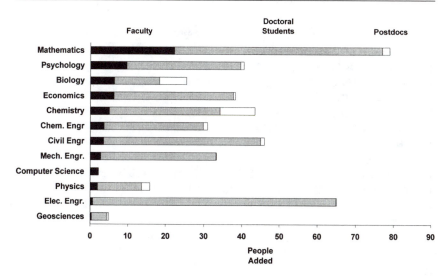

Figure 4.6   Hiring and Admissions Response to $1 Million Increase in Research Funding (All Fields)

A different pattern in hiring and admission decisions emerges across departments in response to an increase in undergraduate nonmajor enrollments. As Figure 4.7 shows, only eight of the 12 fields hire any faculty at all when faced with an additional 1,000 general undergraduates. Two fields have no response whatsoever, indicating that in mechanical and chemical engineering, departments teach only undergraduate majors, not undergraduates from the general student population. In the fields that do increase their human resources to teach general undergraduates, again the bulk of the expansion is in graduate students rather than faculty. In mathematics, which again has the largest response, sizeable numbers of faculty are hired, indicating the strong role mathematics faculty play in teaching undergraduates across all majors, and conversely how much employment is provided for mathematics faculty by general undergraduate teaching demand.

In the final case of demand change when the increased need for teaching comes from more undergraduate majors, every field expands faculty hiring. As Figure 4.8 shows, 10 of the 12 fields also admit more graduate students, who in many cases form the vast majority of the expansion. Note that in many fields, each single additional undergraduate degree in a field induces the typical depart-

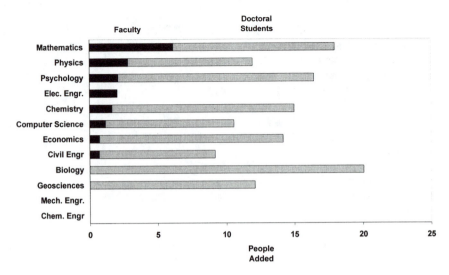

Figure 4.7   Hiring and Admissions Response to 1,000 Increase in General Undergraduates (All Fields)

ment to expand its graduate program by almost one student. (Note that the chart shows the response to 10 new undergraduate degrees, so a value of 10 doctoral students is a one-for-one expansion.) We find this remarkable rate of expansion to be a clear indicator of the internally driven nature of academic department hiring and admissions decisions.

Are faculty substitutes or compliments for doctoral students?

In terms of the production of teaching and research, graduate students and faculty enter into two basic relationships: They act as substitutes for each other and they compliment each other. Both of these relationships exist to some degree or another in all departments across all fields. More elite, research-intensive departments with more graduate students can hire fewer faculty in response to increased teaching demands because the graduate students already on hand can serve as substitute teachers for faculty. The symmetric logic also applies: departments with large faculties relative to teaching demand can admit fewer graduate students when that demand

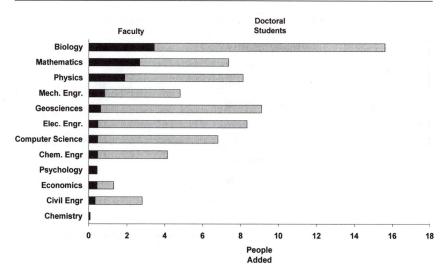

*Figure 4.8   Hiring and Admissions Response to 10 Increase in Under-
graduate Majors (All Fields)*

increases. Conversely, faculty hiring and graduate student admissions can follow closely: More faculty hiring leads to more graduate
student admissions. This complementary type of relationship develops because faculty tend to rely almost exclusively on graduate
students to assist with their research.

Figure 4.9 presents the model results for the more and less elite
segments of the 121 fields with respect to the faculty hiring-doctoral
admissions relationship. The black bars represent the more elite segment in each field, and the white bars are for the less elite segment.
Negative numbers on the sensitivity index indicate that hiring of faculty and admission of graduate students tend to move in opposite
directions, while positive numbers mean they move in the same direction. The model shows that for the more elite, research-intensive
departments, the substitution relationship holds: Having more doctoral students means that fewer faculty are required to meet increases in degree-production and sponsored-research needs. In the
less elite departments, however, the complementary relationship
generally holds: Additional doctoral students are associated with
larger faculty numbers.

These results have important implications for our overall de-

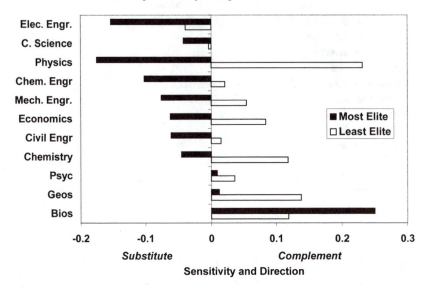

*Figure 4.9   Faculty and Doctoral Students: Substitutes
or Complements*

termination that the US graduate education system is overproducing science and engineering PhDs. While the hypothesis that the production of PhDs is a byproduct of the teaching and research needs in a department is clearly upheld for all departments across all fields, the specific components of these internal department dynamics vary according to academic field and segment type. Generally speaking less elite departments are admitting more PhD students than their more elite counterparts in response to increases in research demand across all academic fields. However, the more elite departments are admitting more PhD students than their less elite counterparts in response to increases in nonmajor teaching. Therefore, in order to bring a halt to the overall overproduction of PhDs, less elite departments should consider alternative ways of addressing their research demand and more elite departments should consider alternative ways of meeting increases in general undergraduate teaching demand.

5. Student Attainment of the PhD

This chapter offers a general summary of our sophisticated modeling of graduate student progress, exploring both the length of time required for students to attain the PhD and the percentage of students who actually earn the degree. In general, the model shows that students today take longer to graduate than in the past, but more students who start graduate programs are completing them. These results are then used to examine the relation between student graduation rates and opportunities for employment outside academia.

Modeling student attainment of the PhD

Figure 5.1 shows a model of students progressing through each year of a graduate program, and either graduating or dropping out. However, calculation of the actual rates of transition from year to year is not possible since comprehensive statistics on the full transition process are unavailable. Indeed much of our strategy in modeling this process involves finding ways around the missing information.

We begin building our model by considering what data are available. Each student earning a doctorate in a US institution is asked to fill out *The Survey of Earned Doctorates*, which has been maintained and compiled by the National Research Council in

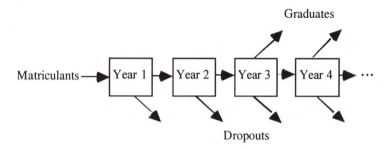

Figure 5.1   Model of Graduate Student Progress

computer databases since the 1960s. In addition, each year the National Science Foundation surveys graduate science and engineering departments to count the total number of enrolled graduate students. Although much more limited, there also are estimates of the total national matriculation by field each year.

Approaching Figure 5.1 from a national level, we have matriculants, current enrollment (that is, the total of cells N1, N2, N3, etc.), and graduations. From *The Survey of Earned Doctorates*, we know the year that each student first enrolled in graduate school and the total time he or she was registered as a student. Since graduate students may delay their education for various reasons, there are two ways to count time in graduate school. One commonly used method simply counts the time from first registration until graduation. However, we use a more complicated method that does not include absences from graduate school, thereby enabling us to compute the total number of dropouts at the national level. In particular, the number of dropouts is the number of graduates minus the number of matriculants minus any increase in the number of students in the pipeline (or plus any decrease in the pipeline). Including these increases and decreases of student numbers across the pipeline allows us to account for students deferring and/or returning to graduate programs at interim points in their education.

We do not automatically know how many students in the pipeline are distributed across each of the cells Year 1, Year 2, Year 3, etc. But by making some assumptions about the stability of the process over a period of time, we can use several years of data to estimate the rates of flow from one year to the next and the rate of graduation or dropout from each cell. Those estimated rates of flow are then used in the main simulation model.

A few additional points should be made with respect to the data collection. First, separate estimations are made for each academic field. Second, because the data allow it and public policy is interested in it, we treat foreign students separately from US citizens and permanent residents. Finally, because all of the data (except the national matriculants series) are available at a disaggregated level, we are able to achieve our ultimate intention—the disaggregation of the national pipeline into separate pipelines for different types of departments. The complete set of methods and assumptions for our model are provided in full detail in Chapter 10.

What the model tells us

Overall, we find that PhD attainment by students entering graduate study in science and engineering averages 23% for US students, 21% for foreign students, and 22% overall. In other words, of every 100 graduate students that enter a graduate department in science and engineering, only about 22 eventually receive a PhD. It is important to note here that the data do not allow us to separate doctoral students from the total pool of graduate students. Many graduate students, especially in engineering, seek only a master's degree and do not intend to pursue a PhD. Conversely, some students who seek only a master's degree actually remain in graduate school to receive a PhD.

PhD attainment is also highly variable by field. As Figure 5.2 illustrates, the science fields tend to have the highest rates of attainment while the engineering fields have the lowest rates. In part this is because there is little motivation to enter science departments with the intent of earning only a master's degree. Some students do stop at the master's level, but usually this is because they lose their desire for a PhD or poor academic performance forces their early withdraw from a PhD program.

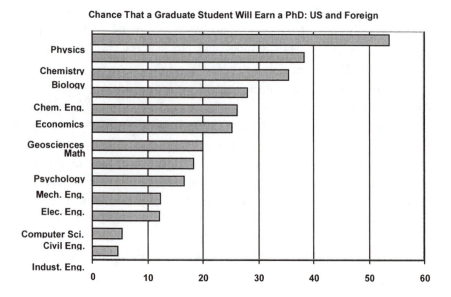

Figure 5.2   Chance that a Graduate Student Will Earn a PhD:
US and Foreign

We also examined the probability of PhD attainment across different segments of the graduate student pool, for example US citizens versus foreigners (in US institutions), students in public versus private graduate institutions, and students in elite versus nonelite graduate institutions. We found no clear trend for either public or private schools to graduate a higher fraction of PhD students. Similarly, there is no clear trend for either elite or nonelite schools to have higher rates of PhD attainment. But when all three factors are examined simultaneously, a trend does appear: Foreigners graduate at lower rates than US citizens in all categories, as shown in Figure 5.3.

With regard to median time to the PhD degree, it is about five years for both US and foreign registered graduate students. But again, there is considerable variability. The model results suggest that there are some "naturals" in the graduate student population who graduate quickly. Indeed, the fastest students attain the PhD in about three years while others can take much longer. The slowest 10% of US students graduate only after more than eight years reg-

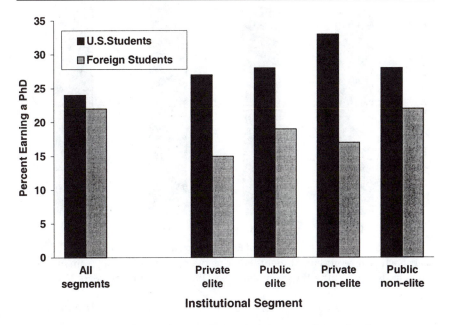

Figure 5.3   Estimated Long-Term Attainment Rates by Segment

istered in graduate school. Very few of the foreign students take this long to graduate, however. Foreigners who are not making satisfactory progress toward the PhD drop out, as evidenced by their lower rates of graduation.

Insights into graduation rates and opportunities for employment

In sum, the model provides several interesting insights into the behavior of students attaining the PhD. As Figure 5.4 illustrates, students today take longer than in the past to receive PhDs; the average is 5.9 years for physics at the end of our data compared to 5.3 years at the start. On the other hand, Figure 5.5 shows that more students who start graduate programs are completing them (though there was a decline in completions in the early 1970s). The higher completion rates are most pronounced in the less elite public institutions.

Interestingly, we also find a relationship between student

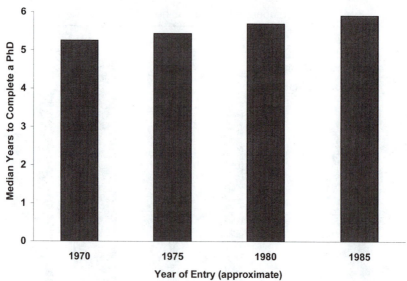

Note: Each year represents students entering over a 10-year period, from five years before to five years after the given date (e.g., 1970 represents the years 1965–75).

Figure 5.4   Registered Time to Degree, Physics and Astronomy,
All Segments

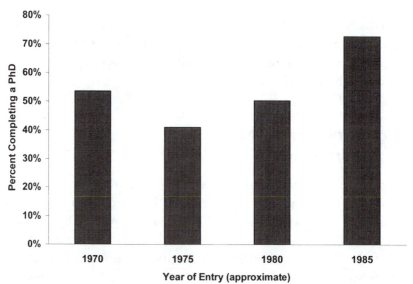

Note: Each year represents students entering over a ten year period, from five years before to five years after the given date (e.g., 1970 represents the years 1965–75).

Figure 5.5   Completion Rates for PhDs, Physics and Astronomy,
All Segments

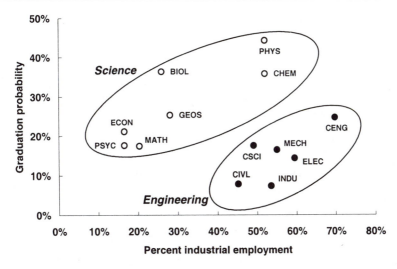

Figure 5.6   Student Graduation Rates Increase with Outside Opportunities for Employment

graduation rates and opportunities for employment outside academia. As highlighted in Figure 5.6, from among the science fields under study, those students with greater employment prospects outside academia have higher PhD graduation rates. The same holds true among the engineering fields. Engineering fields overall, though, have lower graduation rates and higher employment outside academia than science fields. Thus there appears to be more general opportunity for engineering graduate students without PhDs to attain employment in industry. At the same time, greater chances for PhD-level industrial employment also appear to motivate students to attain the PhD. Thus graduate students' perceived opportunities for employment outside academia appear to have a very real motivating effect on their pursuit of both master's and PhD degrees.

6. Faculty Career Patterns

Faculty career patterns, or the rates of hiring, promotion and departure in academic departments, have important effects on the number of new openings for faculty each year. These rates, in turn, have important consequences for the rate of PhD underemployment. In this chapter we present some results from modeling these rates by field and type of institution. In particular, we show that promotion for new professors is most difficult at private liberal arts colleges and private research universities. The chapter also explores some differences in faculty career patterns across the fields of science and engineering.

Modeling faculty career patterns

In order to construct a database of faculty career patterns, we gathered faculty rosters from 10 colleges and universities, spanning the range from liberal arts colleges to major research universities. We obtained rosters for the period between 1968 and 1992 for all the fields under study. Examining these rosters by hand, our research team tracked the progress of each faculty member through assistant, associate, and full professor ranks. In the end, we had data on the number of faculty moving along each path in Figure 6.1, which describes the overall faculty transition system.

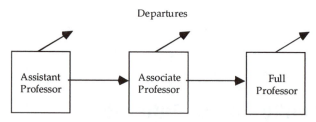

Figure 6.1   Faculty Transition Markov Model

Unfortunately, data on faculty members are not collected systematically at the national level over time. Therefore, in the process of building a model of faculty career patterns, we have to rely on the much more limited set of institution-specific faculty rosters just described. Because these estimates are based on much more limited data than the other portions of our model, we did not insist that our PhD production simulator follow these results strictly. Rather we present a range of estimates that includes these findings as well as an alternative assumption discussed in Chapter 11.

What the model tells us

There are two transitions of special interest to us: promotion rates for new faculty and departure rates for senior faculty. Both of these rates will be important determinants of the number of new openings for faculty each year and, consequently, on the resulting rate of PhD underemployment.

We are interested in the rate at which new faculty are promoted, since promotion generally corresponds to receiving tenure and staying at the institution. Faculty who are not granted tenure must leave their institution. As a result, low rates of promotion increase the shifting of faculty from one institution to another, or the shifting of faculty to nonacademic jobs. However, rates of promotion do not affect the overall balance in the PhD labor market for the field because these relatively new professors will still seek a job.

The model is constructed to estimate the effect of different institutional types and fields on faculty promotion rates. With regard

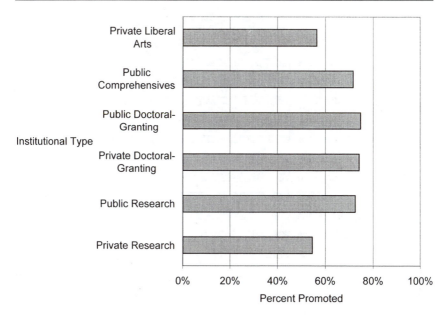

Figure 6.2   Estimates of Promotion Rate for Assistant Professors by Institutional Type

to institutional type, the results presented in Figure 6.2 indicate that promotion for new professors is most difficult at private liberal arts colleges and private research universities. About half of the assistant professors at those schools are promoted within the department, while another half leave the department without promotion. At other types of four-year colleges and universities, almost three-quarters of assistant professors achieve promotion.

There is even more variation in faculty career patterns across disciplinary fields. Promotion rates for assistant professors range from as few as one out four to as many as three out of four, as shown in Figure 6.3. With the exception of mechanical engineering, the highest promotion rates tend to be in the engineering fields. Three fields appear to be clearly tough for new faculty promotions: economics, computer science, and mathematics. Smaller differences across fields should not be given too much weight since the data are insufficient for allowing very precise estimates in such cases.

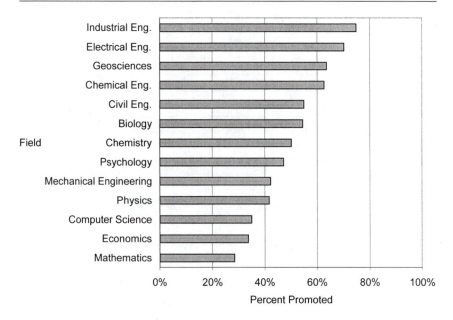

Figure 6.3   Estimates of Promotion Rate for Assistant Professors
by Field

We are most interested in the rate of senior faculty departures. These departures, primarily through faculty retirement, often correspond to a departure from the labor pool, at least on a full-time continuous employment basis. Based on data from the National Survey of Postsecondary Faculty, we calculate that 28%of tenured faculty who leave their departments take a new job and 72% permanently exit the labor force through retirement, death, or disability. These calculations are described in more detail in Chapter 11.

Figure 6.4 shows the results of our model for the rate at which full professors depart from their departments each year. The average rate across all fields is about 4% per year, but the range is substantial even within disciplines. Among the engineering fields, for example, rates of senior faculty departure are as low as 2% to 3% for mechanical, electrical, and chemical engineering and as high as 6% for industrial engineering.

In sum, our model shows that promotion rates for new faculty

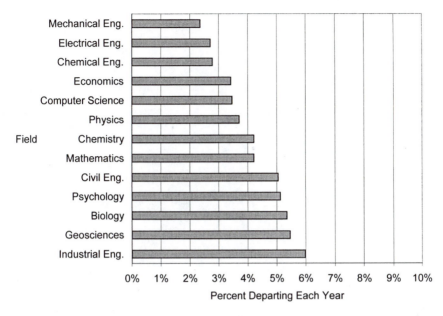

Figure 6.4   Estimates of Departure Rates for Full Professors by Field

vary widely across disciplinary fields, just as departure rates for senior faculty range widely within disciplines. The one definitive finding of our model is that new faculty at private liberal arts colleges and private research universities have the most difficult time getting promoted.

7. Conclusions About the Academic System

In this concluding chapter of Part I, we attempt to shed light on the entire system of higher education and the wider social systems that affect and are affected by it. We do so by drawing together important insights offered in the previous chapters.

Based on the faculty and department chair interviews reported in Chapter 2, we hypothesize that science and engineering departments see PhD students as labor to meet the teaching and research demands of their departments. The results of the departmental decision-making model in Chapter 4 indicate that most fields do in fact admit PhD students to serve teaching and research needs, although the specifics of these internal department dynamics vary significantly by department. Generally speaking, across all academic fields, less elite departments admit more PhD students than their more elite counterparts in response to increases in research demand. However, the more elite departments admit more PhD students than their less elite counterparts in response to increases in nonmajor teaching. Therefore, in order to bring a halt to the overall overproduction of PhDs, less elite departments should consider alternative ways of addressing their research demand, and more elite departments should consider alternative ways of meeting increases in general undergraduate teaching demand.

Faculty, of course, also serve teaching and research needs in science and engineering departments. There are reasons for concern in this area as well. In other work, Massy and Wilger (1995) have documented an effect they call the academic ratchet, whereby faculty shift their time increasingly toward research and away from teaching. The academic ratchet is a further symptom indicative of an academic system in that is responsive to internal, rather than external, imperatives.

In Chapter 5, we found that PhD attainment by students entering graduate study in science and engineering averages 24% for US students, 22% for foreign students, and 23% overall. In other words, of every 100 graduate students that enter a graduate department in science and engineering, only about 23 eventually receive a PhD. Unfortunately, it is very difficult to place this 23% attainment rate in context. As noted throughout this work, we cannot say how many of every 100 graduate students are truly pursuing a PhD. Some who intended to get a PhD may stop at the master's degree. Others intending for a master's may persist to attain a PhD. Even so, high rates of PhD attainment do not appear to be a major cause of overproduction of PhDs. If departments are producing more PhDs than the economy can absorb, it is not because they fail to weed out many students. Three-quarters of science and engineering graduate students never receive a PhD. The one-quarter who do receive the PhD enter the PhD labor market.

Th detailed estimations summarized in Chapter 4 indicate that students in different groups exhibit significant differences in their behavior. These results validate the need to group departments according to major categories of research intensity. Although it is clear that foreign students face different incentives in graduate study, we must take departmental differences in research into account before we can observe different degree attainment for foreign students.

Separating departments into more and less elite confirms that foreign students are much less likely to persist to obtain a PhD in each type of department and institution. For example, in the most elite, research intensive, private departments, 27% of US students attain the PhD, but only 15% of foreign students do. The overall figures mask the large differences in attainment for US and foreign students.

Foreign students do indeed appear to be responding to the incentive to use US graduate schools as an entry path to the US labor market. US students already have eligibility to work in the US and need not attend graduate school for that reason. Therefore it is understandable that US students who do go to graduate school go with more intention of attaining the PhD. Some foreign students, on the other hand, may be in US graduate schools in order to gain access to employment in the US and not because they have a strong need to get a PhD.

The volume of students admitted to graduate programs in science and engineering combined with attainment rates for the PhD degree produce the annual number of new PhDs each year. This is one side of the PhD labor market. The other side is the creation of employment positions for PhDs in academia, government, industry, and overseas. The model presented in Chapter 3 shows that in almost all the fields we studied, in a steady state, the creation of positions lags behind the production of new PhDs. The principal drivers within academe are demand for education by undergraduate and master's level students and the volume of sponsored research.

Of course, science and engineering fields are not necessarily in a steady state. If education demand or sponsored research increases steadily and quickly enough, the PhD employment system comes into balance. But if growth levels off, the system starts spilling out PhDs without creating opportunities for them to do what they are trained to do in graduate school.

If fluctuations are short-term, the postdoc pool can absorb the differences between supply and demand. When the system is expanding rapidly, postdoc pools can shrink. When the system stops expanding, postdoc pools grow again. If the system stops expanding for too long, though, postdoc pools keep growing, eventually pushing more and more scientists out of science careers as permanent employment prospects worsen.

There are several possible responses to this diagnosis. Graduate schools might adjust their admissions and hiring practices. Graduate schools might alter the training they provide so that graduates are better prepared for a wider variety of careers.

Prescriptions on foreign students necessitate hard choices. It is clear that the US benefits from highly capable and highly trained scientists moving here from around the world to invent new technologies. But current immigration policy is at odds with the academic reality. Immigration policy expects foreign students to return to their home countries within a short time after earning a degree. Yet the public subsidy of graduate education implies that the recipients of that subsidy remain in this country to make use of their training in ways that benefit the US economy. Allowing departments open access to accept foreign students permits the departments to insulate themselves from information about the domestic demand for their graduates. Even if US students are discouraged by the prospects in a field, the chance to study for any science PhD at a US university remains so attractive to foreign students that we expect a ready supply of potential applicants from overseas. Because much of the costs of science training are borne by state and federal taxpayers, these students do not need the means to pay for their training. Whatever their intentions, many of these students stay in the US for some time.

On the basis of the work reported in this volume, we believe that US immigration policy should recognize that graduate students commonly remain in the US after graduation, whether at the master's or doctoral level. We also believe it is appropriate to place some restrictions on the number of foreign graduate students admitted to science departments.

Broadening science training beyond conventional research content sounds like a friendly option, but it has several flaws as a remedy on its own. First, faculty face strong incentives to hire graduate students to further their own research careers. They will likely resist attempts to broaden PhD training beyond the conduct of research. And from a larger social perspective, they may be wise in a certain sense, because the second flaw in this remedy is that the Federal and state governments pay for the costs of conducting research and hence subsidize the training of science and engineering PhDs. In the face of this subsidy, training more PhDs with broader career aspirations may be a better deal for the students, but it is not likely to be a better deal for the taxpayers. The students will get broader exposure to different aspects of research and its applications, but this is a singularly expensive way to train generalists who will not perform research themselves.

This analysis pushes us to advocate academic restructuring, including a re-examination of the way academic teaching and research are performed. Each department is a relatively small part of the total training and employment system. Therefore it is tempting for each department to say, "Let others change their admissions practices or the ways they conduct teaching and research, while we continue as before." When every department, or most departments, reason that way, nothing changes overall.

Now, departments and individual scientists are largely free to decide who teaches courses and who performs work on research grants and contracts. If the academy does not restructure internally, Federal and state policymakers may find opportunities to modify the system. State policymakers have considered and, in some cases, already imposed limitations for public universities on the number of credit hours that departments can use graduate students as opposed to faculty members for teaching. Federal policymakers might restrict the unfettered ability of scientists to hire as many graduate students as their grants can support.

The independence of the academy has many benefits for intellectual inquiry and the advance of knowledge. Without undermining these benefits, the academy must find a way to respect the interests of the participants in the system: undergraduate students, graduate students, postdocs, faculty. Beyond the interests of those in the system, we believe it is appropriate to remember that much of the costs of academic science are borne by the general public. Based on the analysis in this volume, we believe that the public and certainly the student participants in the system would be well served by some restructuring of graduate education.

References

Massy, W. F., & Wilger, A. (1995, July–August). Faculty productivity. *Change.*

Part II

Details of the Modeling

8. Modeling the Employment Gap for PhD Scientists and Engineers

The study covers virtually all doctoral and four-year nondoctoral-granting colleges and universities in the United States. The only institutions excluded were those which did not respond to the requisite government surveys and those where we encountered insurmountable missing-data problems in the course of analysis. Our final database included some 210 doctoral-granting and more than 1,000 nondoctoral-granting institutions.

Most seriously, basic data on the number of faculty in each discipline at each institution are lacking at the national level. There is no frequent survey of faculty numbers by field and institution. About once a decade, the National Research Council (NRC) assembles several measures in each field for many doctoral-granting departments. These measures including faculty size. At the time of our work, the 1980 data were available, so we used those data. Because industrial engineering is not included in the NRC data, we had no data on faculty size. That field is omitted from the overall simulation model, although it is included in some of the estimations of the models for doctoral student progress and faculty careers in Chapters 10 and 11.

Data on other variables come mainly from the NRC, the National Science Foundation, and the National Center for Educational Statistics. Most of the data were extracted from the NSF's CASPAR database. The NRC provided special tabulations on time to degree by institutional segment from its Survey of Earned Doctorates. Data on faculty transitions were extracted from various college catalogues. More information on specific data is provided in later chapters.

We decided early in our analysis that the standard institutional categories (e.g., the Carnegie classifications crossed with type of control) would be too coarse for modeling the doctoral-granting sector. Hence we devised a method for partitioning the public and private doctoral-granting institutions into relatively homogeneous groups ("segments"). The methodology uses information about the scale of activity in each field in each institution to group similar scales of activity together into segments. Results of this procedure are documented in Appendix A.

Model structure

The model consists of three major components: the Departmental Choice Model; the Faculty Transitions Model; and the Doctoral Production Model. Two smaller models, labeled "submodels" in the figure, also enter the simulation. Starting with four "driver" or exogenous variables and measures of the labor demand from non-US academic institutions, the simulation computes a net employment gap, which represents PhD underemployment.

Driver variables

We simulate the actions of academic departments and their students as they pertain to doctoral production and utilization. The simulation begins with the four exogenous ("driver") variables shown at the top of Figure 3.1.

BachAll: the number of bachelors degrees awarded by the institution, a surrogate for total undergraduate enrollment and hence the department's general-education teaching load

Bach: the number of bachelors degrees awarded by the department (i.e., in the department's discipline), a surrogate for the department's in-major teaching load

Master's: the number of master's degrees awarded by the department, a surrogate for the department's taught-postgraduate teaching load

R&D: the amount of sponsored research funding in the department's discipline, a surrogate for departmental research activity

The departmental choice model

The output of the departmental choice model consists of three variables: 1) the total number of departmental tenure-line faculty; 2) the total number of doctoral students enrolled by the department; and 3) the number of postdoctoral appointees in the department.

The faculty transitions model

The faculty transition analysis estimates entry to the faculty and exit from the faculty of a single department, but we still must account for the entries and exits from the labor market as a whole. The fraction of departures from individual departments that represent departures from the labor force is based on data from the 1988 National Survey of Postsecondary Faculty (NSOPF-88). Over a one-year period from 1986 to 1987, of all tenured faculty who left their departments, 2.7% retired, about 0.4% died or became disabled, and 3.2% sought a new job (source: Russell et al., 1991, p. 20 for retirements and p. 31 for other departures). Thus, 28% of all tenured departures sought a new job and remained active in the labor force. Based on this parameter, the simulation counts 28% of the departures by full professors as remaining in the active labor force. All departures at the level of assistant or associate professor are retained in the active labor force.

Because the NSOPF data are known to be problematic, we recomputed the model results using an alternative assumption about faculty transitions. In the alternative, we use a constant rate of retirement and replacement for faculty, equal to the rate used for US nonacademic sectors: 2.75% per year.

While on the subject of employment transitions, let us consider the simple system used to model postdoctoral appointments. We assume that each appointment lasts for exactly one year, after which the person returns to the labor pool. Putting postdocs in the model makes no difference to the supply-demand equilibrium, but it does affect the system's response to changes in the driver variables.

The doctoral production model

The stochastic model provides transition probabilities for each year from one to has high as needed to account for time in program. (We used a 20-year cutoff.) It also provides the unconditional probabilities of graduation and dropout for each year, the asymptotic graduation probability, and the expected time to degree for various percentiles of the time-to-degree distribution.

The simulation system

The simulation system brings the various models together to project, for all US four-year colleges and universities, the supply of new doctorates and returnees to the doctorate labor pool, and the demand for faculty and postdoctoral appointments. By adding the rates of employment outside academia, including overseas demand, and domestic industry and government demand, the system can be used to generate the long-term supply-demand equilibrium for a given drives configuration or to compare the equilibria generated by different driver configurations.

The projections should be viewed in a comparative statics sense because the model was not tuned to show the time path traversed in moving from one equilibrium to another. The simulation does generate a path between starting values and the steady-state, but the response to driver fluctuations seems faster than would be expected in practice. The departmental choice model allows departments to adjust fully from their current status to any new desired point each year, subject to the constraints of not dismissing faculty or graduate students except through the ordinary processes of tenure decisions, departures, and retirement.

In practice, departments do not adjust their employment and resource use so fast. Hence the models transition paths are shorter and less smooth than expected in the real system. The model is adaptable to such smoothing and lengthening of transitions, but such a refinement is unnecessary to illustrate the characteristics of the academic production system.

Despite the simulation system's richness of detail, the basic model boils down to a large first-order Markov process. The state-space includes the numbers of US and foreign doctoral students by program stage, faculty numbers by rank, and the number of post-doctoral appointments—all on a segment-by-segment basis. The model is Markovian because its behavior depends only on the current state variables, not on the path by which the current state was reached. It is Markovian steady state to which we refer when describing "long-run equilibrium." We obtain the equilibrium figures by simply letting the model run until the requisite time-invariance is obtained.

Specifically, we start the model with initial conditions data for 1980, the first year of our data series. Next we calculate the time trend for each driver's evolution from 1980 to 1991 (1990 for R&D), and then let the model run to equilibrium with the drivers equal to their 1992 projected trend value. This equilibrium generates the data used to compute the employment gap in each field.

Computing the employment gap

The model describes inflows and outflows from the PhD labor market from the academic sector: new PhDs produced, postdoctoral appointments, new faculty hired, and faculty returning the labor market. To complete the labor market picture, we include two more sectors: all other domestic US demand, and foreign demand. Domestic demand includes PhD-level positions in colleges and universities that are not regular faculty or postdoctoral appointments as well as all demand from industry, government, and nonprofit organizations.

While data on the inflows and outflows by field do not exist, we do have information on total doctorate employment in the domestic noneducational sectors. These data can be combined with information produced by the model to make rough approximations about underemployment. We defined a "base employment gap" by assuming that nonacademic demand remains constant at the replacement rates needed to maintain the early 1990s employment levels. Another way of stating this assumption is that we fix the appropriate industrial (and governmental) employment level for PhDs in each field at its 1990s level.

Future industrial employment increases may or may not reduce the "underemployment" of PhDs as we have defined the term. Underemployment would decrease if and only if the new jobs represent cost-effective use of PhD training. Cost-effectiveness refers to a world without subsidies, and we know of no way to model or make assumptions about such a world. To avoid making quantifying assumptions in advance, we use the familiar strategy of setting a straw man—in this case the breakeven growth rate for nonacademic employment—and making a binary judgment about whether each field's breakeven rate is reasonably possible. Of course our readers are free to make their own binary judgments. Atkinson (1990) reports that, for all science and engineering fields, about half of foreigners receiving PhDs in the United States remain in this country. Based on these results, we assumed that 50% of foreigners (those on temporary visas) would continue to remain in the US labor market.

To compute the employment gap, we take the annual number of PhDs produced, add faculty and postdocs returning to the labor market, subtract new faculty hires and postdoctoral appointments, and subtract demand from abroad, and from US nonacademic sectors. To calculate the breakeven growth rates of nonacademic employment, we divide the base employment gap by the estimated value of such employment (using the model's estimates nonacademic employment rather than the data from SESTAT in order to maintain internal consistency). These calculations were illustrated in Chapter 3 for the field of mathematics. Here we present more detailed results for all of the 12 fields we studied.

Results for all fields

The results for mathematics are representative of the results for most of the other science and engineering fields we studied. Results for all fields are summarized in Tables 8.1 through 8.3, positioned at the end of the chapter. Table 8.1 describes the static employment gap, while Tables 8.2 and 8.3 describe the system's dynamic response to increases in sponsored research and degree output (all sub-doctoral degrees), respectively.

Table 8.1 Employment Gap

	PHYS	CHEM	BIOS	GEOS	MATH	ECON	PSYC	CENG	CIVL	ELEC	MECH	CSCI	Total
Simulated data (base equilibrium)													
Total faculty	10,198	12,017	38,537	4,539	19,011	10,067	17,080	1,654	3,141	6,048	4,441	7,148	133,880
Retirements	242	317	679	133	372	242	483	32	103	165	110	189	3,067
Retirement rate	0.024	0.026	0.018	0.029	0.020	0.024	0.028	0.019	0.033	0.027	0.025	0.026	0.023
Career length	42	38	57	34	51	42	35	52	31	37	40	38	496
New doctoral degrees	1,126	1,744	2,855	728	1,073	989	1,915	508	501	1,288	791	920	14,439
Percent degrees by US citizens	73.4%	44.5%	82.9%	47.1%	65.6%	53.6%	96.7%	66.8%	47.4%	55.0%	49.1%	32.7%	64.9%
NSF doctorate employment data													
Total employed doctorates	31,905	48,967	66,037	13,263	20,049	19,241	65,672	10,633	7,512	16,994	8,680	21,000	329,953
Implied nonacademic employment	21,708	36,950	27,499	8,724	1,038	9,174	48,592	8,979	4,371	10,946	4,239	13,852	196,072
Employment gap estimates:													
Academic and nonacademic retirement rate = 0.0275													
Academic retirements	280	330	1,060	125	523	277	470	45	86	166	122	197	3,682
Nonacademic retirements	597	1,016	756	240	29	252	1,336	247	120	301	117	381	5,392
Foreign Demand	150	484	245	193	184	230	32	84	132	290	201	309	2,534
Estimated US employment gap	99	-87	795	170	338	230	77	132	163	531	351	33	2,832
% of new degrees	8.8%	-5.0%	27.8%	23.4%	31.5%	23.3%	4.0%	25.9%	32.5%	41.2%	44.4%	3.6%	19.6%
"Breakeven" growth rate of nonacademic employment	0.5%	-0.2%	2.9%	2.0%	32.5%	2.5%	0.2%	1.5%	3.7%	4.8%	8.3%	0.2%	1.4%
Simulated retirement for academic, nonacademic retirement rate = 0.0275													
Academic retirements	242	317	679	133	372	242	483	32	103	165	110	189	3,067
Nonacademic retirements	597	1,016	756	240	29	252	1,336	247	120	301	117	381	5,392
Foreign Demand	150	484	245	193	184	230	32	84	132	290	201	309	2,534
Estimated US employment gap	137	-73	1,175	162	489	265	63	145	146	532	363	41	3,447
% of new degrees	12.2%	-4.2%	41.2%	22.3%	45.5%	26.8%	3.3%	28.5%	29.2%	41.3%	45.9%	4.5%	23.9%
"Breakeven" growth rate of nonacademic employment	0.6%	-0.2%	4.3%	1.9%	47.1%	2.9%	0.1%	1.6%	3.3%	4.9%	8.6%	0.3%	1.8%

Table 8.2 Sponsored Research Dynamics

	PHYS	CHEM	BIOS	GEOS	MATH	ECON	PSYC	CENG	CIVL	ELEC	MECH	CSCI
Base employment gap	137	−73	1176	162	489	265	63	145	146	532	363	41
as percent of degrees	12%	−4%	41%	22%	46%	27%	3%	29%	29%	41%	46%	4%
During expansion												
Employment gap	67	−190	819	149	439	249	31	135	134	552	357	29
as percent of degrees	6%	−11%	28%	20%	41%	25%	2%	26%	27%	43%	44%	3%
Elasticities												
Faculty hires	1.08	0.99	1.52	0.51	0.68	0.39	0.36	1.81	0.65	−0.48	0.80	0.36
Doctoral admits	0.56	0.70	0.79	0.18	0.45	0.26	0.24	0.56	0.48	0.60	0.49	0.00
Faculty	0.20	0.21	0.28	0.09	0.10	0.06	0.07	0.20	0.12	−0.08	0.12	0.06
Doctoral students	0.37	0.51	0.32	0.08	0.25	0.15	0.14	0.34	0.30	0.40	0.28	0.00
Postdocs	0.70	0.88	0.80	0.45	0.74	0.39	0.34	0.56	0.58	0.40	0.18	0.09
Employment gap	−5.09	15.99	−3.04	−0.84	−1.03	−0.62	−5.08	−0.66	−0.85	0.37	−0.16	−2.93
Emp. gap/degrees	−5.11	15.78	−3.08	−0.85	−1.07	−0.65	−5.09	−0.74	−0.88	0.35	−0.31	−2.94
After expansion												
Employment gap	131	−98	1261	167	511	277	92	163	155	583	383	40
as percent of degrees	11%	−5%	42%	23%	46%	28%	5%	31%	30%	43%	47%	4%
Elasticities												
Faculty hires	0.20	0.20	0.34	0.11	0.13	0.08	0.08	0.25	0.15	−0.10	0.14	0.08
Doctoral admits	0.38	0.48	−0.15	0.12	0.33	0.18	0.18	0.41	0.38	0.52	0.34	0.00
Faculty	0.20	0.21	0.35	0.11	0.12	0.08	0.08	0.26	0.15	−0.10	0.15	0.08
Doctoral students	0.37	0.51	0.40	0.11	0.31	0.18	0.17	0.43	0.38	0.51	0.36	0.00
Postdocs	0.70	0.88	1.01	0.57	0.94	0.49	0.42	0.70	0.73	0.51	0.23	0.11
Employment gap	−0.41	3.41	0.72	0.27	0.47	0.46	4.46	1.20	0.62	0.95	0.55	−0.25
Emp. gap/degrees	−0.755	2.749	0.300	0.161	0.158	0.272	4.220	0.744	0.315	0.438	0.181	−0.252
Change												
Change in Employment Gap	6	25	−85	−4	−23	−12	−28	−17	−9	−51	−20	1

Table 8.3 Degree Production Dynamics

	PHYS	CHEM	BIOS	GEOS	MATH	ECON	PSYC	CENG	CIVL	ELEC	MECH	CSCI
Base employment gap	137	-73	1176	162	489	265	-310	145	146	532	363	41
as percent of degrees	12%	-4%	41%	22%	46%	27%	-1%	29%	29%	41%	46%	4%
During expansion												
Employment gap	-36	-231	891	85	156	125	-655	95	78	353	252	-82
as percent of degrees	-3%	-13%	31%	12%	14%	13%	-3%	19%	16%	27%	31%	-9%
Elasticities: "During"												
Faculty hires	3.92	3.07	2.76	3.96	4.66	3.37	3.96	10.39	4.33	7.48	6.76	3.67
Doctoral admits	0.79	0.59	1.72	0.99	0.69	0.73	0.39	0.24	0.48	0.65	0.64	1.32
Faculty	0.59	0.49	0.50	0.66	0.66	0.53	0.71	1.17	0.79	1.12	0.97	0.64
Doctoral students	0.44	0.34	0.69	0.54	0.42	0.41	0.23	0.15	0.32	6.44	0.35	0.70
Postdocs	0.33	0.00	0.00	0.00	0.00	0.00	0.35	0.00	0.00	0.00	0.00	0.36
Employment gap	-12.66	21.58	-2.42	-4.79	-6.80	-5.29	11.11	-3.49	-4.65	-3.37	-3.07	-29.99
Emp. gap/degrees	-12.64	21.37	-2.54	-4.84	-6.83	-5.33	10.64	-3.51	-4.67	-3.42	-3.15	-29.88
After expansion												
Employment gap	129	-94	1366	187	508	289	-305	149	148	562	375	82
as percent of degrees	11%	-5%	44%	24%	45%	28%	-1%	29%	29%	41%	46%	8%
Elasticities												
Faculty hires	0.75	0.62	0.62	0.86	0.86	0.67	0.91	1.43	1.02	1.38	1.20	0.82
Doctoral admits	0.54	0.40	0.84	0.64	0.50	0.50	0.28	0.18	0.37	0.46	0.46	0.89
Faculty	0.75	0.63	0.63	0.84	0.83	0.67	0.90	1.48	0.99	1.41	1.23	0.81
Doctoral students	0.55	0.43	0.87	0.68	0.53	0.52	0.29	0.19	0.40	6.59	0.44	0.89
Postdocs	0.42	0.00	0.00	0.00	0.00	0.00	0.44	0.00	0.00	0.00	0.00	0.46
Employment gap	-0.59	2.85	1.61	1.51	0.40	0.89	-0.18	0.27	0.14	0.55	0.34	9.80
Emp. gap/degrees	-1.078	2.321	0.672	0.793	-0.125	0.366	-0.451	0.074	-0.190	0.031	-0.077	8.175
Change												
Change in Employment Gap	8	21	-190	-25	-20	-24	-5	-4	-2	-29	-12	-40

Base employment gap

In 10 of the 12 fields we studied, our simulations resulted in large base employment gaps (over 10% of annual production). Two fields (psychology and computer science) showed small gaps (less than 10%) and one field (chemistry) showed no gap at all. These results come from independent simulations. Each field's simulation depends on data unique to the field. The consistency of the results is therefore noteworthy.

In aggregate the base employment gap for these fields is between 20% and 24% of annual PhD production. If the premises of the simulation are correct and the resources available to academic science and engineering stabilize at current levels, each year one-fifth or more of new PhDs will be underemployed unless cost-effective nonacademic employment can take up the slack. The breakeven growth rates range from nil to the range of 33% to 47% per year. Chapter 3 discusses these rates and compares them to actual data on post baccalaureate science and engineering employment. We argue there that the breakeven growth rates needed to close the employment gaps in the majority of fields will turn out to be unsustainable.

Sensitivity to driver changes

What if resources do not stabilize at current levels? As in the case of mathematics, increases in sponsored research or degree production that stabilize at a new higher level generally result in a worsened employment gap.

The tables begin with the base employment gap, in absolute terms and as a percent of doctoral degrees granted. Then we show the employment gap in the fifth year of the expansion period ("during expansion"), followed by elasticities for the simulation's key variables. This tableau is repeated at the bottom of the table for the long-run equilibrium results ("after expansion"). As described earlier, the drivers are assumed to grow at 2% per year for five years.

In every case, the increased resources increase employment in the short-run. Examining the figures reported in Tables 8.2 and 8.3 for the employment gap during expansion confirms this observation. In physics and computer science the increase in degree production eliminates the employment gap completely.

The elasticities shown in the tables allow us to examine the mechanisms that operate during the expansion. Increased sponsored research leads to more faculty hiring and more postdoctoral appointments, as indicated by the positive elasticities for faculty fires and postdocs (except for the single unusual case of faculty hiring in Electrical Engineering). When degree production is increased rather than sponsored research, faculty hiring is affected even more strongly and uniformly for all fields. Postdoctoral appointments are not sensitive to increases in degree production in nine fields, with small increases in three fields. The departmental behavior in most fields implies that postdoctoral appointments are determined exclusively, or primarily, by sponsored research.

These increases in faculty and postdoctoral employment create many new job openings, reducing the employment gap. But there is another side to the expansion. Departments begin to admit more doctoral students when resources increase. In most fields, the increases in graduate programs are larger when degree production rises than when sponsored research rises.

Some of these new doctoral admits will eventually graduate with PhDs. As the resources again stabilize, these new PhDs crowd the job market. Considering the 11 fields that have an employment gap in the simulation (all but chemistry), there are 22 dynamic simulations: one each for sponsored research and degree production. Out of those 22 simulations, 14 result in sizeable increases in the employment gap (more than 10 positions per year change). Eight simulations result in small changes (less than 10 positions increase or decrease). None of the 22 simulations suggest that these fields are self-regulating, meaning that higher levels of resources do not tend to close the employment gap.

Labor market feedback

The model and results described here assume that departments and faculty do not adjust their behavior in response to conditions in the labor market for the PhDs they produce. In particular, we assume that the types of people employed in teaching and research do not vary and the intake of graduate students does not vary. These assumptions are justified by the structure of the academic department, as described in Chapter 2.

To a great extent, departmental resistance to using labor market feedback stems from a valid concern that data on labor markets, especially forecasts of future demand, are notoriously shaky. The model of this book, indeed, is not intended to be a forecast of actual labor market experience.

Faculty claim that because the lead time for producing PhDs is so long and forecasts of future demand are so hazy, they are obligated to continue their production unmodified. But the aggregate consequences of such atomistic behavior could well be problematic.

Feedback models

While most of our analysis uses an open-loop system, we did formulate and test two additional submodels describing possible negative effects of supply-demand imbalance upon doctoral admissions. We describe the models as operating on the doctoral applicant pool, but one of them also could represent the impact of supply-demand imbalance upon departmental choice.

Theory suggests that potential students' willingness to pursue doctoral study depends in some way on the strength of the employment market for doctorate-holders. In terms of our simulation, the relevant assumption would be that application demand depends inversely on the employment gap—our measure of supply-demand imbalance. The larger the employment gap, the fewer the total number of doctoral applications, other things being equal. The linkage would gain strength if large employment gaps depress faculty salaries (and conversely), or if the employment gap is highly visible and the danger of underemployment highly salient to prospective candidates. We are not convinced that either condition represents reality, but that need not deter us from modeling "what if" questions about the sensitivity of applicant pools to the doctoral labor market.

We explored several methods of incorporating feedback and the resulting need to ration graduate students across departments. Since the market is limiting the availability of potential graduate students, the departmental resource allocation model cannot operate unconstrained. At least some departments must be restricted from admitting doctoral students and substitute other human resources, mostly faculty.

More work remains to be done in the area of feedback and adjustments. We remain convinced that neither departments nor prospective doctoral students take close account of the doctorate employment gap in any case.

Conclusion

The results of our simulations have convinced us that both of the supply-demand imbalance hypotheses are credible. While we cannot prove that such imbalances exist, our painstaking efforts to model PhD production and faculty demand have projected base employment gaps in 11 of the 12 cases for which we could obtain the requisite data. Of the 11 fields with base employment gaps, 10 have long-term dynamic imbalances based on changes in sponsored research or degrees. In a majority of fields, the growth rates of industrial and government demand that would be needed to close the employment gaps appear unsustainable. The results for different fields rest on entirely separate datasets, so they can be regarded as largely independent replications. The case for imbalance in the science and engineering PhD system is thus a strong one.

References

Atkinson, R. C. (1990, April 27). Supply and demand for scientists and engineers: A national crisis in the making, *Science, 248,* 425–432.

Russell, S. H., et al. (1991). *1988 National Survey of Postsecondary Faculty (NSOPF-88): Profiles of faculty in higher education institutions,* NCES 91-389, U. S. Department of Education, Office of Educational Research and Improvement, National Center for Education Statistics.

9. Modeling Production of Teaching and Research in Departments

This chapter describes a model for how human resource levels are set at the departmental level. Taking departmental workloads (non-doctorate degree production and sponsored research) as drivers, the model predicts departmental demand for doctoral students, faculty, and postdoctoral fellows. All three endogenous variables contribute to degree and sponsored research production; faculty also contribute to the production of doctorates.

The model is rooted in the microeconomic theory of nonprofit enterprises, wherein decision makers allocate resources by maximizing a utility function subject to production and financial constraints (for reference, please see Hopkins & Massy, 1981; James, 1982, pp. 350–366; and James, 1990, pp. 77–106). Departments may attribute intrinsic utility to the endogenous variables and also value them instrumentally because of their contributions to production. We model and estimate both kinds of utility effects as well as the financial effect. Using these estimates, we predict the endogenous variables by means of mathematical programming. Estimated income, price, and transformation elasticities can then be obtained by varying the corresponding parameters and drivers.

The chapter consists of three sections. The first section describes the model's mathematical structure and develops the needed parameter estimation procedure. The second section analyzes the physics field in considerable detail: first in terms of the meaning of the coefficients, then in terms of the estimator's stability, and finally in terms of the sensitivity of the departmental simulator to changes in the driver variables' prices and quantities. The third section extends the physics results to all 12 science and engineering fields included in our dataset. This produces the parameters used in the doctorate supply-demand simulation reported in Chapter 8.

The model

Data and definitions

Our database consists of records for the science and engineering departments of each institution included in the National Science Foundation's CASPAR files. While much government data are available as time series, the key figures needed for this study are limited to a single cross-sectional sample for the year 1980. For each department, we define:

Endogenous variables

Fac: faculty (FTE)[1]
Doc: doctoral enrollments (headcount)
PostDoc: postdoctoral fellows (FTE)

Exogenous variables

BachAll: undergraduate enrollment for the institution as a whole (FTE)
Bach: baccalaureate degrees granted in the major (number)
Master's: master's degrees granted in the major (number)
R&D: total sponsored research generated by the department ($000)

Departmental descriptors

S-index: segment index (a continuous variable, defined below, where the positive direction indicates "elite" institutions); in equations, *S-index* is abbreviated to *S*.

Toc: dummy variable for type of institutional control (0 for public institutions; 1 for private institutions); in equations, *Toc* is abbreviated to *T*.

For compactness, we sometimes denote the endogenous variables by the vector $\mathbf{y}$, the exogenous variables by $\mathbf{x}$, and the segment descriptors by $\mathbf{z}$.

We obtained the segment index from the analysis used to divide doctorate-granting departments into homogeneous groups for purposes of simulation (see Appendix A). For physics, this produced four private-sector segments and six public-sector ones. The value of *S-index* for each segment was set to equal the average factor score for the segment, with the more "elite" institutions having the larger scores. Under this scheme, *S-index*=0 represents the hypothetical "average institution."

Three additional segments represent the nondoctoral-granting institutions: private liberal arts colleges, private comprehensive universities, and public colleges and comprehensive universities. The original *S-index* and *Toc* are irrelevant for these segments (as are *Doc* and *PostDoc*), but we added a new segment variable based on the institution's Carnegie classification: *S-index*=1 for LAC1 and C1, and 0 for LAC2 and C2. In contrast to the doctoral sector, where the data for all the segments were pooled, the model is estimated separately for each nondoctoral segment.

Structural equations

Our presentation begins with the financial constraint and then moves on to the utility and production functions. Parameter estimation procedures are developed in the following section, and then statistical results will be reported.

Finance. The financial constraint prevents the total cost of *Doc*, *Fac*, and *PostDoc* from exceeding the department's resources as determined by the exogenous variables. The equation involves three types of coefficients:

- $\mathbf{c}$, the department's *marginal costs* for *Doc*, *Fac*, and *PostDoc*, taken as negative to denote cost.

- **r**, the department's *marginal revenues* for *BachAll, Bach, Master's, and R&D*, as these revenues are applied to defray the costs of *Doc, Fac,* and *PostDoc,*
- r_0, *fixed revenue minus fixed cost*, where "fixed" means not associated with *Doc, Fac,* and *PostDoc.*

Using these definitions, we write the financial constraint as:

$$\mathbf{c'y} + r_0 + \mathbf{r'x} \geq 0 \qquad (9.1)$$

Doctoral tuition and financial aid are excluded from both sides. These quantities would offset each other because nearly all doctoral students receive financial support.

The revenue assumptions are most easily justified when an explicit budgeting formula allocates unrestricted funds to departments in proportion to enrollment and sponsored research income. We need not assume formulaic budgeting, however, since any resource allocation system will respond eventually to exogenous revenue shifts.

The so-called revenue theory of budgeting holds that universities raise all the money they can and spend all the money they raise (Bowen, 1980), and we believe the same applies to academic departments. This means the financial constraint binds on average, so (9.1) can be written as an equality. We can estimate each department's **c** from independent data as described below; hence we collapse $\mathbf{c'y}$ to a single number and move it to the left-hand side-which produces a regression equation. We also postulate that r_0 and **r** interact linearly with the segment descriptors, which leads us to define interaction functions as illustrated here for revenue in the kth segment:

$$r_k = r_{k\cdot} + r_{ks}S + r_{kT}T \qquad (9.2)$$

Henceforth, all references to r_0 and **r** should be understood to include the possibility of segment interactions. Equation (9.4), presented later, shows the resulting financial regression.

Estimation of marginal cost begins with Association of University Personnel Administrators' data on overall average faculty compensation, and average assistant professor compensation, for all institutions and fields. (The compensation figures include fringe benefits.) The next step adjusts the overall average by the CASPAR figure for average faculty compensation for the institution in question. The final step, which produces a field and institution-specific index, multiplies the institution's average by the ratio of faculty salaries in the field being analyzed to the average for all fields (College and University Personnel Association, 1982,1987, and 1993). Informal evidence suggests that doctoral TAs and RAs receive about one-third the salary of an assistant professor, and that postdoctoral fellows receive about two-thirds of an assistant professor's pay, so we used those ratios. Fortunately, tests showed the model to be insensitive to the exact figure.

Production and utility.   We assume that departments value all three endogenous variables. Doctoral and postdoctoral students are valued as outputs of the educational process. Faculty are valued not only as inputs but also, as in the case of monks in the medieval monastery, "for the good of the scholarly order"—that is, for their own sake, as colleagues. The main value effect, the linear term in the utility function, is represented by the vector product $\lambda'\mathbf{y}$, where the elements of λ are all positive. As in the case of the financial coefficients, we assume that depends on the segment characteristics according to the scheme described in (9.2).

Departments also recognize that *Doc*, *Fac*, and *PostDoc* represent inputs to the educational and research processes. Resources must be spent on them in order to produce *BachAll*, *Bach*, *Master's*, and *R&D*, and to obtain the revenues associated with these outputs. The simplest formulation holds that the three inputs must be present in sufficient quantities to produce the exogenous output quantities at historic quality levels. Following this line of reasoning, we postulate the following linear relationships:

$Doc \geq L_D[BachAll, Bach, R&D]$

> A department's doctoral degrees depend on institution-wide and in-major undergraduate degrees because of the need for TAs, and on sponsored research because of the need for RAs.

Fac≥ L$_F$[BachAll,Bach,Master's,Doc,R&D]

Faculty size depends on general undergraduate degrees, in-major degrees, master's degrees, and doctoral degrees because of instructional needs; and on sponsored research because of the need for research assistants.

PostDoc≤L$_P$[Fac,R&D]

The number of postdoctoral students is limited by the number of faculty supervisors and sponsored research funding.

We assume that the coefficients in the constraints depend on *S-index* and *Toc*, just as in (9.2).

According to the above formulation, the department's decisions would maximize $\lambda'\mathbf{y}$ subject to the production and financial constraints. Suppose, however, a department views the constraints not as strict requirements, but rather as norms or rules of thumb. Violations are penalized in the utility function, because they threaten output quality or financial viability. Slack is penalized as well, but for a different reason: the penalty for slack represents diminished marginal utility. Suppose, for example, that a department considers the possibility of dramatically increasing faculty size while holding doctoral and postdoctoral enrollments constant at their norms. Penalizing utility for the resulting faculty-constraint slack increases the marginal rate of substitution (in utility terms) of *Fac* for *Doc* and *Fac* for *PostDoc*. If we assume that both penalties are quadratic, the new formulation satisfies the basic assumption required for optimization—namely, that the objective function should be convex.

It seemed likely, given the above logic, that the penalties for violating the production constraints would be greater than those associated with slack. Therefore, we make provision for asymmetric penalties by writing the basic model in the form of a goal program (Hillier & Lieberman, 1990, p. 271). Writing positive and negative deviations and their associated weights with the superscripts "+" and "−", the model becomes:

$$u[\mathbf{y}] = \lambda'\mathbf{y} - \sum_{k=1}^{3} w_k^+ \{(\mathbf{y} - L_k[\mathbf{y},\mathbf{x}])^+\}^2$$

$$- \sum_{k=1}^{3} w_k^- \{(\mathbf{y} - L_k(\mathbf{y},\mathbf{x}))^-\}^2 - \{(\mathbf{c}'\mathbf{y} + r_0 + \mathbf{r}'\mathbf{x})^-\}^2 \quad (9.3)$$

which is to be maximized with respect to **y** for **y**≥0. The financial constraint remains an inequality, with only negative deviations entering the objective, since we assume no penalty for running a budget surplus. (The positive utility weights on **y** preclude a substantial surplus, however.) The financial-function weight has been pegged at one without loss of generality, which makes it the system's numeraire.

Estimation

We estimate the parameters of (9.3) from cross-sectional data. Estimation proceeds in two stages. First we estimate the financial coefficients by ordinary least squares; then, we estimate the production coefficients by a special procedure to be described below. We did not attempt to estimate the penalty asymmetries directly. Instead, we estimated the parameters of a symmetric approximation and calculated the asymmetry factor from the residuals.

We define the following extensions to the exogenous variables to handle interactions with the segment descriptors during the estimation process.

$$\mathbf{y}_{z,i} = \begin{bmatrix} \mathbf{y}_i \\ z_{1i}\mathbf{y}_i \\ z_{2i}\mathbf{y}_i \end{bmatrix}, \quad \mathbf{x}_{z,i} = \begin{bmatrix} \mathbf{x}_i \\ z_{1i}\mathbf{x}_i \\ z_{2i}\mathbf{x}_i \end{bmatrix}$$

where $x_{z,i}$ and $y_{z,i}$ are column vectors of 9 and 12 elements, respectively, representing the ith department in our sample.

The financial equation.   The marginal costs for *Doc*, *Fac*, and *Post-Doc* have been calculated from exogenous financial data as described above, so we have the coefficients in **c'y**. We form the following regression in the unknown r_0 and **r**:

$$\mathbf{c'y}_i = y_i^* = r_0 + \mathbf{r'}\begin{bmatrix} \mathbf{x}_{z,i} \\ \mathbf{z}_i \end{bmatrix} + u_i \tag{9.4}$$

where y_i^* is the dependent variable, [·] is a 14-element column vector joining $x_{z,i}$ and z_i, and **r** the corresponding vector of revenue coefficients. The coefficients (**r**) and regression constant (r_0) can be estimated by ordinary least squares. The residuals can be viewed as transitory surpluses and deficits—deviations from financial equilibrium that will be corrected by future decisions. We will designate the negative residuals by u_i^-, where $u_i^- = 0$ if the ith residual is positive.

Departmental decision rule. The symmetric departmental objective function can now be written as:

$$\max_{y_i} 2\lambda' y_i - (\Gamma y_{z,i} + B x_{z,i})' W (\Gamma y_{z,i} + B x_{z,i})$$
$$- \{(c' y_i + r_0 + r' x_{z,i})^-\}^2 \qquad (9.5)$$

where Γ and B are the production-function coefficients, W is a diagonal matrix of weights, and the last quantity equals u_i^-. The necessary conditions for maximizing (9.5) can be written as:

$$\lambda_{k0} + \lambda_{kS} S_i + \lambda_{kT} T_i - w_k y_k - w_k M_{k,i} f_k[\theta] - c_k u_i^- = \varepsilon_{k,i},$$
$$k = 1,2,3 \qquad (9.6)$$

M_{ki} is a vector of polynomials in x_i, y_i, and z_i, and $f_k[\theta]$ is a vector of polynomials in the elements of Γ, B, and W. (The definitions of M_{ki} and $f_k[\theta]$ are given in Appendix Table A-3.) The right-hand side ($\varepsilon_{k,i}$) is a stochastic variable representing error in the department's optimization. We assume that the expectation of $\varepsilon_{k,i}$ is zero.

Estimation procedure. Estimates of the financial-function coefficients are already in hand, so our task is to estimate the elements of λ, Γ, B, and W. The system is triangular, with the third equation (for *PostDoc*) standing alone, the second (for *Fac*) depending on the third, and the first (for *Doc*) depending on the other two. Because the system is triangular, it meets all the conditions for identifiability, and we can estimate the equations sequentially: (9.3) first, then (9.2), and finally (9.1).

Estimation for the kth equation begins by regressing y_k on $x_{z,i}$ and z_i to obtain the reduced-form coefficients. The reduced-form predictions are used everywhere except for the left-hand sides of the structural regressions (shown below), where actual values are used; this removes the structural errors from the right-hand-side y-variates, thus avoiding bias. The next step is to multiply the equation by $1/w_k$ (which we define as ϕ_k) and bring y_k to the left-hand side, which produces the regression:

$$y_k = \phi_k \lambda_{k0} + \phi_k \lambda_{kS} S_i + \phi_k \lambda_{kT} T_i - M_{k,i} f_k[\theta] - \phi_k c_k u_i^- + \varepsilon_{k,i},$$
$$k = 1,2,3 \qquad (9.7)$$

This step is required for y_k to serve as the dependent variable: that is, for the sum of squares to be minimized in the y_k direction, consistent with the purpose of predicting y_k, rather than in direction of the financial-equation residuals. We then proceed as follows to estimate the parameters of (9.7) for the kth equation. (Step iii does not apply to the nondoctoral segments, since *Fac* is the only endogenous variable.)

i. Subtract off the means of **M,** S,-*index Toc,* and $c_k u^-$, which removes λ_{k0} from the equation.

ii. Minimize the sums of squared errors of the three equations in (9.6) with respect to θ_k and ϕ_k.

iii. Where the equation depends on the prior equations' θ, we proceed iteratively to make the w-values from the successive equations consistent with one another:

   a. Set $w_k = 1$ and estimate the equation; if ϕ_k equals the ϕ from the previous equation, proceed directly to step iv.

   b. Update w_k by multiplying the previous one by the ratio of the newly estimated ϕ_k to the ϕ estimated in the previous equation.

   c. Repeat step "b" until the estimated ϕ_k become stationary.

iv. Obtain λ_{k0} from (9.7) by inserting the estimated value of θ_k and adding back the means of all the variables.

v. Calculate the coefficients' standard errors by multiplying the inverse of the information matrix (the matrix of objective-function second-order partial derivatives, evaluated at the estimated parameter values) by twice the equation's variance of estimate.

vi. When all the equations have been estimated, divide λ and **w** by the ϕ-value as estimated from the last equation (equation 1), to obtain the final λ and **w**.

The above amounts to a two-stage least squares procedure, so the estimates have the usual desirable properties. We have not tried to prove convergence of the ϕ-iteration, but as a practical matter the procedure appears robust.

The $f_k[\theta]$-functions (defined in connection with 9.6) are nonlinear in the parameters, so the normal least-squares solution procedure cannot be used. Therefore, we adopted a Gaussian optimization procedure (Beightler, Phillips, & Wilde, 1979, p. 228). The method works well when the objective is the square of an underlying nonlinear function, in our case a polynomial in the unknown parameters. The iteration usually converges in only a few iterations.

Examination of the residuals indicated that, as expected, their distributions are skewed to the right for all three production equations and also for the financial equation. This indicates that the penalties for violating constraints are greater than the ones for slack, but the question of quantifying the production-asymmetry remains. We proceed by partitioning the residuals of (9.7) into positive and negative components and writing the weights for the partitions as $m_k^+ w_k$ and $(1-m_k^+) w_k$. The necessary conditions for a maximum require that the expected penalties for positive and negative penalties be equal at the department's optimum decision point:

$$E[\{m_k^+ w_k \mathbf{M}_{k,i} f_k[\theta]\}^2] - E[\{(1 - m_k^+)w_k \mathbf{M}_{k,i} f_k[\theta]\}^2] = 0 \quad (9.8)$$

It is a simple matter to numerically minimize the sum of squares corresponding to (9.8) with respect to m_k^+.

Simulation

Departmental choice behavior can be simulated by applying (9.3) to the exogenous variables. The simulation uses aggregate values for *BachAll, Bach, Master's,* and *R&D,* treating the segment as if it were a single institution, and produces aggregate values for *Doc, Fac,* and *PostDoc.*

Calibration.   Since the optimization is not a linear operation, applying (9.3) to an aggregate **x** does not necessarily reproduce the segment's aggregate **y**. Hence we added a segment-specific calibration constant to each constraint, so that each holds exactly at the observed aggregate **x**. The calibrations assume that each aggregate "department" operates at its production norms and on its budget constraint, and that it will set *Doc, Fac,* and *PostDoc* to equal the observed **y** when presented with the segment's observed **x**. Once calibrated, the model predicts the "department's" responses to variations of x from its observed value.

Limit constraints.   Additional constraints are needed when we use the departmental decision simulator in our larger model. Suppose, for example, that the simulator calls for fewer faculty than currently reside on the department's roster and that the model's attrition expectations do not produce the desired faculty numbers even if hiring is brought to zero. Simply setting faculty to the achievable number is not appropriate, since the doctoral and postdoctoral population would be no longer consistent with faculty size. Hence we include limiting constraints in the optimization itself. In the above example, the lower limit on faculty would equal the expected population with no hiring. Absent constraining circumstances, the lower bounds are set to zero and the upper ones to plus infinity.

Piecewise-linear approximation.   The optimization is best accomplished through quadratic programming. Lacking a suitable quadratic programming routine callable from Mathematica, we used a piecewise-linear function to approximate the quadratic objective in a linear programming formulation (Hillier & Lieberman, 1990, p. 528). Each of the seven terms in (9.3) was approximated by $h+1$ linear segments. The first h pieces span distances equal to the square root of sum of squares of D^+ or D^-, according to the term being calculated; the last piece in each set is unbounded. Experimentation indicated that $h=10$ is sufficient to capture the quadratic's curvature for the range of deviations encountered in the simulation.

Binding financial constraint.   Estimation proceeded under the assumption that departments would usually spend all available funds: hence only the negative financial residuals were used in equations (9.5) and (9.6). Tests with the simulation, however, showed that our estimates of the weighting factors (w_k) are not precise enough to allow the equations to be used in their original form. Specifically, the weights appear to be too large. The production constraints overpower both the utility coefficients and financial constraints. Income improvements fail to call forth additional faculty, doctoral students, and postdocs; income reductions fail to limit departmental decisions, which leads to uncontrolled deficits.

To solve this problem, we modified the simulation algorithm to require the financial constraint to bind at all times. That is, we made the financial constraint two-sided and gave it essentially an infinite weight. The utility coefficients and production-constraint weights continue to function in terms of relative importance, but their role in determining overall resource use has been preempted by the always-binding financial constraint.

Results for physics

This section explores the estimator's properties and demonstrates the decision simulator by using data for a single field: physics. Results for all twelve science and engineering fields included in our dataset are presented afterward.

Both the financial-function and production-function estimators performed well. No problems were observed with production-function iteration convergence or intermediate results. The signs of nearly all the coefficients are consistent with our theory, most of the coefficients are statistically significant, and the R-squares are generally quite high.

Financial equations

Financial equations: doctoral segments. Table 9.1 reports three versions of the financial equation for the doctoral segments. The first column shows the base case, the one we used in the simulation: we will describe these results shortly. *Test A* adds physics master's degrees to the equation. The *Master's* coefficient is very large and highly significant statistically. The figure is so large, $63,000 per master's degree, that we did not find it credible. Adding *Master's* also exerts a powerful effect on the *BachMajor* coefficients. The main effect is cut in half, and the sign of the interaction with type of control (*toc*) is reversed. The equation's degree of fit remains constant, however.

We believe that most physics master's degrees are granted as milestones on the way to the doctorate or as consolation prizes,

Table 9.1   Financial Regressions: Doctoral Segments

	Base run	Test A	Test B		Base run	Test A	Test B
Fixed revenue	-194.1 (108.5)	-30.6 (94.1)	-149.8 (87.7)	R&D	0.227 (0.028)	0.189 (0.024)	0.263 (0.029)
BachAll	0.236 (0.041)	0.194 (0.035)	0.204 (0.039)	R&D*S-index	-0.253 (0.058)	-0.212 (0.049)	-0.225 (0.062)
BachAll*Toc	-0.147 (0.103)	-0.122 (0.087)	-0.018 (0.079)	R&D*Toc	0.052 (0.042)	0.104 (0.036)	0.025 (0.045)
BachMajor	11.592 (5.638)	6.259 (4.825)	20.588 (5.783)	S-index	1763.2 (335.9)	272.7 (333.1)	—
BachMajor*Toc	9.276 (11.50)	-3.894 (9.882)	14.689 (12.24)	Toc	216.5 (172.6)	254.0 (146.5)	—
Master's	—	63.04 (7.29)	—	R-squared	83.1%	83.1%	72.3%

which implies a definitional relation between *Master's* and the number of doctoral students. The latter, one will recall, enters into the dependent variable for the financial equation. According to this line of reasoning, the effects of *Master's* are spuriously large, and the variable should be excluded from the equation.

Test B deletes the segment variables, S-index and toc, from the base-case model. R-squared drops by almost 11 points, but the only other material effect occurs for BachMajor. Both the main effect and the interaction with toc increase by about 50 percent; however, the relation between them is virtually unchanged. Adjusting fixed revenue according to the segment descriptors removes some of the effect that otherwise would be attributed to in-major degrees, a result we consider credible in view of the size of the coefficients and the strong statistical significant of S-index in the base model.

Turning now to the base case, in the first column of Table 9.1, we see that the main effects for the three included x-variates (Bach-All, Bach, and R&D) are positive, and all are statistically significant. Next come the interactions with type of control. BachAll*Toc is negative, but since it is less than BachAll the combined effect when Toc=1 is still positive. None of the Toc interactions are significant, but we included them because we seek the best possible point estimates for purposes of simulation. R&D*S-index is negative, larger in absolute value that its main effect, and highly significant. The largest S-index value (0.8) occurs when Toc=1, however, so the combined effect of the three variables remains positive. The coefficient for the public segment with the largest S-index value (0.46) is positive as well. The main-effect coefficient for S-index is highly significant while that for Toc is not, but both exert a material effect on fixed revenue.

The coefficients present a fresh view of how funding for physics academic salaries depends upon output levels. For example, every additional general bachelor's degree adds $235 to the average physics budget in public institutions but only $89 to private institutions. Apparently the public sector's greater reliance on formula funding percolates all the way to the departmental level. Both types of institutions respond strongly to variations in the number of physics baccalaureates: the budget increments are $11,590 in the public sector and an even more generous $20,870 in the private sector. (The figures refer to degrees attained, not to enrollments; the latter would produce considerably smaller marginal revenue estimates.)

The average private department has positive fixed revenue, since the positive Toc coefficient is larger than the negative intercept term. The more elite the institution, the larger the fixed revenue. Departments in the lowest S-index category have negative fixed revenue—that is, they are burdened with fixed costs that must be covered from their incremental revenues. Positive fixed revenue occurs most often in institutions with substantial endowment (Hopkins & Massy, 1981, Chapter 3), which is likeliest in the most elite private segment.

In the area of R&D, only 22.7% of the incremental dollar covers departmental academic salaries in public departments versus 27.9% in their private counterparts. These figures do not seem unduly low: in addition to the academic salaries defined in **c** (and hence included in the above percentages), R&D funding must cover other research expenses like technicians, travel, and equipment, and also university overhead. We do not know why the Toc coefficient is positive—a negative coefficient would have been consistent with the private sector's higher overhead rates—but it is not close to being statistically significant. Within each sector, departments in the more elite institutions receive substantially less funding than those in less elite institutions: this is consistent with the latter's proclivity for rebating overhead back to the department and principal investigator.

Our results can also be described by the financial tradeoffs between pairs of variables shown in Table 9.2:

Table 9.2   Breakeven Values

Type of control:	Private	Public
Bach:BachAll (%)	0.4%	2.0%
Bach:BachAll (n:n)	237	49
R&D:BachAll ($:n)	$321	$966
R&D:Bach($:n)	$74k	$51k

For public institutions, the model indicates that physics majors produce the same departmental marginal revenue as overall baccalaureate production does if the former represents 2% of total student numbers (*BachAll/Bach*=0.02). The breakeven figure is only 0.4% for private institutions, indicating greater sensitivity to majors in the private sector. The relation is illustrated further in the table's next line, which shows that private institutions require 237 total degrees to produce the departmental revenue associated with one physics major, as compared to only 49 degrees in the public sector. Physics departments in private institutions must get $321 in extra R&D support to equal the amount obtained from one general baccalaureate degree—a figure that triples in the public sector. On the other hand, the more lucrative funding for majors in private departments reverses the terms of trade when R&D is compared to in-major degrees.

Financial equation: nondoctoral segments. Table 9.3 presents the financial-equation coefficients for the nondoctoral segments: since the nondoctoral equations are relatively uncomplicated, only the base runs are presented. Unlike the doctoral segments, which are pooled for purposes of estimation, the three nondoctoral segments are estimated separately. The model is somewhat different as well, due to the different segment designators, response sensitivities, and

Table 9.3 Financial Regressions: Nondoctoral Segments

	Segment 11: private liberal arts	Segment 12: private comprehensive	Segment 13: public lib. arts & comprehensive
Fixed revenue (cost)	6.0	−0.3	−7.2
	(3.4)	(6.4)	(12.5)
BachAll	0.145	0.136	0.170
	(0.172)	(0.017)	(0.010)
BachMajor	3.253	9.187	4.198
	(0.869)	(1.680)	(1.894)
Master's	104.98	28.786	
	(17.290)	(8.342)	
R&D	0.428	0.289	0.068
	(0.036)	(0.011)	(0.056)
S-index	40.3	12.7	
	(4.7)	(8.2)	
R-squared	61.1%	85.7%	73.1%

collinearity patterns. Master's degrees were included because there is no possibility of confounding with (nonexistent) doctoral enrollments. The degree of fit exceeds 60% in all three segments.

Private liberal arts college departments garner less revenue per degree from in-major baccalaureate enrollments, and more from master's degrees, than do those in private comprehensive institutions. Revenue per degree from aggregate enrollment is about equal for the two private segments. The public institutions' undergraduate revenue-per-degree figures are similar to the privates', but the result for master's degrees was small and statistically insignificant. As might be expected, R&D is most important in the public sector, moderately important for the private comprehensives, and essentially zero for the private liberal arts colleges.

S-index is positive for both private segments but essentially zero for public institutions. It is largest in the liberal arts colleges, which indicates that the Carnegie-I colleges (where *S-index* equals one) tend to maintain physics department budgets independently of enrollment fluctuations to a greater extent than their less elite (and less affluent) counterparts. The negative fixed-revenue and zero *S-index* coefficient in the public sector imply that departments must garner enrollments to balance their budgets. The same is true for the nonelite private liberal arts colleges, for which the value of *S-index* is zero.

Production equations

Production equations: doctoral segments.   Tables 9.4 through 9.6 present the production-estimation results for the doctoral segments—again for the base run used in the simulation and two alternatives. Table 9.4 gives the results for the Doctoral equation, Table 9.5 for the Faculty equation, and Table 9.6 for the Postdoctoral equation. The figures for *BachAll* and *Weight* have been multiplied by 1000 and those for *Bach* and *Master's* by 10 to improve interpretability. The estimator performed well; no problems were observed with iteration convergence or intermediate results. We will describe the base run presently, but first, as in the case of the financial equation, we will discuss the two test runs.

Test A follows up on our concern about including *Master's* in the financial equation and also examines the effect of removing *Fac-*

Table 9.4 Financial Regressions: Nondoctoral Segments

	Base run	Test A	Test B		Base run	Test A	Test B
Utility	1.766	1.749	3.798	Bach	6.273	6.257	7.273
	(0.235)	(0.235)	(0.561)	(0)	(1.521)	(1.529)	(1.871)
Utility*S-index	15.974	15.771	—	Master's	—	—	—
	(2.620)	(2.638)		(0)			
Doc	—	—	—	Master's*Toc			
				(0)			
Doc*S-index	—	—	—	Master's*Toc			
Fac	—	—	—	R&D	11.63	11.60	11.30
				(000)	(0.83)	(0.80)	(1.00)
				R&D*S-index	−9.32	−9.40	−7.10
					(1.20)	(1.20)	(1.50)
BachAll	9.098	9.100	7.300	Weight (000)	0.0724	0.0970	0.1627
(000)	(1.036)	(1.000)	(1.200)		(0.0122)	(0.0145)	(0.0070)
BachAll*S-index	31.89	32.10	47.90	R-squared	91.7%	91.6%	87.1%
(000)	(4.32)	(4.30)	(3.9)	df	193	193	194

Table 9.5   Faculty Production Coefficients: Doctoral Segments

	Base run	Test A	Test B
Utility	12.47	12.27	23.87
	(1.78)	(1.48)	(3.5)
Utility*S-index	51.97	54.38	—
	(2.62)	(19.85)	
Doc	0.184	0.202	0.179
	(0.053)	(0.027)	(0.051)
Doc*S-index	-0.449	-0.386	-0.392
	(0.049)	(0.046)	(0.050)
Fac	—	—	—
BachAll (000)	2.847	2.500	2.700
	(0.641)	(0.500)	(0.700)
BachAll*S-index (000)	—	—	—
Bach (0)	1.876	1.759	1.867
	(0.650)	(0.600)	(0.697)
Master's (0)	0.559	—	2.012
	(2.106)		(1.872)
Master's*Toc (0)	0.764	—	1.233
	(1.426)		(1.479)
R&D (000)	2.013	1.700	2.100
	(0.386)	(0.300)	(0.400)
R&D*S-index Weight (000)	1.730	1.667	2.564
	(0.206)	(0.300)	(0.510)
R-squared	77.6%	77.5%	72.8
df	191	193	192

Table 9.6　Postdoctoral Production Coefficients: Doctoral Segments

	Base run	Test A	Test B		Base run	Test A	Test B
Utility	0.650 (0.974)	4.112 (0.469)	-2.467 (2.274)	Bach (0)	—	—	—
Utility*S-index	37.218 (5.819)	42.736 (5.375)	—	Master's (0)	—	—	—
Doc	—	—	—	Master's*Toc (0)	—	—	—
Doc*S-index	—	—	—	R&D (000)	2.31 (0.45)	3.80 (0.30)	1.80 (0.50)
Fac	0.303 (0.067)	—	0.450 (0.074)	R&D*S-index (000)	-1.09 (0.56)	-2.80 (0.40)	0.20 (0.60)
BachAll (000)	—	—	—	Weight (000)	0.735 (0.098)	0.735 (0.097)	1.545 (0.042)
BachAll*S-index (000)	—	—	—	R-squared	78.6%	76.3	71.2
				df	195	196	197

ulty from the postdoctoral equation. (Additional tests deleted the variable sets one at a time; the results did not differ materially from those reported here.) We discussed the *Master's* problem in connection with the Financial equation. The sensitivity of *Faculty* was tested because it does not enter the Postdoctoral equation for most of the other fields.

We were particularly interested in the effect of removing *Master's* on the *Doctoral* coefficient in the faculty equation, even though it was not statistically significant. The Financial equation proved to be unstable when *Master's* was included, and one might have expected the same for the Faculty equation due to the correlation between faculty and doctoral-student numbers. The two-stage least-squares procedure, however, has taken care of the problem. Neither the doctoral coefficients nor their standard errors are materially affected by the change. This result gains importance because of the interesting interpretations for the substitutability of doctoral students for faculty in the elite segments, which will be discussed below.

Removing *Faculty* from the Postdoctoral equation increases both the utility and the R&D coefficients. The main utility effect rises dramatically and its standard error drops by half. Hence it appears that, for physics, taking account of faculty reduces the apparent valuation of Postdocs. In other words, with Faculty absent from the equation, some of the value associated with faculty spills over to Postdocs. This effect may also be operating in the two other fields where faculty entered the Postdoctoral equation, psychology and computer science, since the ratio of the *PD* to *Fac* utility coefficients are rather small for those fields as well. The effect may operate in reverse in the other 9 fields, where *Faculty* did not enter the Postdoctoral equation because of insignificance or, in a few cases, collinearity problems. It does not seem likely that an insignificant *Fac[PD]* coefficient would affect the utility coefficient materially, however. The R&D coefficient also rise in absolute value, but the effect is not nearly as pronounced as the one for utility.

Test B deletes *utility*S-index* from all three equations. The resulting utility main effects change materially, and the Postdoctoral coefficient goes negative. The other coefficients do not change much, but the R-squares drop by about five percentage points. We concluded from the test that including *utility*S-index* is worthwhile because it improves the utility estimate and enhances fit without affecting the rest of the structure.

We turn now to the base run, which provides some interesting insights into the world of academic physics. Looking first at the main-effect utility coefficients (*utility*), we see that the value placed on an extra faculty member is much greater than that for doctoral students and postdocs, and that the *Fac* and *Doc* coefficients are highly significant. These coefficients refer to the average segment, for which *S-index*=0. (*Toc* was dropped from all three equations because of collinearity problems.)

The value placed on the three endogenous variables increases as one moves to the more elite segments, with the increase being largest for faculty and smallest for doctoral students. For the most elite segment (*S-index*=0.8), faculty are valued at 54 compared to postdocs at 30.5 and doctoral students at 14.6. For the least elite segment (*S-index*=–0.1), the figures are 7.3 for faculty, 0.2 for doctoral students, and –4.6 for postdocs. Departmental decisions depend on the ratios rather than the raw values of the utility coefficients. These ratios are shown in Table 9.7.

The negative excursion of postdoctoral utility in the least elite segment probably stems from the small numbers of postdocs relative to faculty and doctoral students in those institutions. For purposes of simulation, we simply clipped the negative excursion at zero—making the lowest segment indifferent to the use of postdocs as shown in the table. We do feel confident, however, in concluding that the more elite segments value postdocs more strongly relative to faculty and doctoral students than does the average segment. The more elite segments also value doctoral students more highly relative to faculty than the average segment does.

Table 9.7   Ratios Affecting Departmental Decisions

	U[Doc]/U[[Fac]	U[PD]/U[[Fac]	U[PD]/U[[Doc]
Average segment	0.14	0.05	0.37
Most elite segment	1.17	2.57	12.12
Least elite segment	0.01	0	0

The differences between more and less elite departments are treated in Chapter 4.

Production equations: nondoctoral segments. Parameter estimates for faculty, the only equation that applies to the nondoctoral segments, are shown in Table 9.8. We varied the equation by adding the nondoctoral segment descriptor (*S-index*), but it did not produce worthwhile or statistically significant effects. As for the doctoral segments, the equations are well behaved and explain more than 75% of the variance in faculty size.

The direct utility coefficients (*utility*) are positive, and two of the three are statistically significant. The public institutions appear to value faculty most strongly, followed by the private liberal arts colleges and comprehensives in that order. We should note, though, that these coefficients don't normally affect the simulation because—with only one decision variable—no tradeoffs come into play.

The effects of general and physics degrees are highly significant in all three nondoctoral segments. The private institutions seem to be more sensitive than their public counterparts, with the liberal arts colleges responding somewhat more to general degrees, and the comprehensives to degrees in the major. General bachelor's

Table 9.8   Production Coefficients: Nondoctoral Segments

	Seg. 11 pvt. LAC	Seg. 12 pvt. Comp.	Seg. 13 public
Utility	0.313	0.056	0.578
	(0.100)	(0.136)	(0.207)
BachAll	8.354	7. 639	6.596
(000)	(0.417)	(0.343)	(0.238)
Bach	2.073	3.494	1.077
(0)	(0.212)	(0.389)	(0.480)
Master's	7.585	12.601	—
(0)	(4.444)	(1.930)	
R&D	17.208	11.466	34.190
(000)	(1.093)	(0.251)	(1.441)
R-squared	78.8	95.2%	86.1%
df	351	293	298

degrees exert a much stronger effect on faculty in the nondoctoral segments than in the doctoral-granting segments. Probably the latter's larger size dilutes the impact of general enrollment variations on individual departments. The in-major effects fluctuate across the nondoctoral segments, but the average is comparable to the average doctoral segment.

Master's degrees drive faculty size to a far greater extent in the private nondoctoral segments than in the doctoral sector. For private liberal arts colleges, an additional ten Master's degrees calls forth 7.6 faculty, whereas for private comprehensive universities, the ratio is greater than 12.6 to 10. The *Master's* coefficient for the public comprehensives was negative but less than its standard error, so we deleted it.

The effect of R&D on faculty size also is much larger in the nondoctoral institutions than in the doctoral-granting ones. In private liberal arts colleges, an extra million dollars of departmental R&D "buys" 17 additional faculty; the figure falls to about 11 faculty in the private comprehensives, but it jumped to 34 faculty in the public nondoctoral sector. The larger nondoctoral effect comports with the proposition that doctoral-granting departments must spread research funds among doctoral students, postdoctoral fellows, and faculty, whereas the nondoctoral departments can concentrate them on faculty.

Asymmetry in the residuals

We expected the residuals for both the finance and production equations to be skewed to the right: i.e., that departments view slack in the constraints less seriously than violations. The rightward skewness does in fact occur in every case—in all the equations for both the doctoral and the nondoctoral segments.

Simulation sensitivity analysis

What do the statistical results imply about departmental choice behavior? No single set of coefficients can provide the information needed to answer this question. All the equations interact, and the result can only be obtained by actually performing the optimization set forth in equation (9.3).

While the graphs in Chapter 4 provide a useful overview of the model's sensitivity, they do not provide accurate quantification. Hence we calculated a "sensitivity elasticity" for each pair of variables—the percentage change in the decision variables per one-percent change in the independent variable. The sensitivity elasticities are presented in Table 9.9.

Doctoral and postdoctoral enrollments are elastic with respect to fixed revenue (their elasticities are greater than one in absolute value). Departments add most to doctoral enrollment when revenue grows, with the percentage of faculty additions being less than half that for doctorates. All the other variables are inelastic. Nondoctoral departments add faculty at a greater rate than doctoral-granting ones, but the greater cost of faculty keeps the size of the increment relatively low. The nondoctoral departments are relatively sensitive to the price of faculty (c_F in the revenue function) and to the prices and quantities of general baccalaureate degrees. Faculty size in the doctorate segments depends most heavily on *BachAll*, and then on *Price:BachAll* and *Price:R&D*. Doctoral enrollment depends most heavily on the price of faculty—perhaps largely due to an income effect, since faculty account for the largest part of the department's

Table 9.9 Simulation Elasticities (percent change in column variable per one percent change in row variable)

	Doctoral			Nondoctoral
	Doc	*Fac*	*PostDoc*	*Fac*
Fixed revenue	0.202	0.023	1.036	0.788
Price: *Doc*	−0.309	−0.061	−1.086	—
Price: *Fac*	−0.577	−0.169	−1.678	−0.804
Price: *PostDoc*	−0.098	−0.019	−0.647	—
Price: *BachAll*	0.392	0.131	1.338	0.638
Price: *Bach*	0.116	0.031	0.611	0.066
Price: *Master's*	0.000	0.000	0.000	0.006
Price: *R&D*	0.269	0.059	1.054	0.068
Quantity: *BachAll*	0.416	0.421	0.191	0.638
Quantity: *Bach*	0.125	0.121	0.138	0.066
Quantity: *Master's*	−0.015	0.019	−0.069	0.006
Quantity: *R&D*	0.270	0.221	0.339	0.068

salary budget. Next comes the price of doctorates themselves, the price and quantity of general bachelor's degrees, and the amount of R&D. Postdoctoral enrollment depends mainly on R&D quantity, and then on the prices of faculty, doctoral students, and R&D.

Results for all fields

Having completed our in-depth analysis of physics, we now turn our attention to parameter comparisons across all 12 of the scientific and engineering fields included in our dataset. Key differences among the fields are discussed in Chapter 4. Tables 9.10 through 9.13 summarize the estimation results for both the doctoral and nondoctoral segments. The model continued to work well when extended beyond the physics dataset. The R-squares generally exceed 50%, and most of the coefficients are statistically significant.

Table 9.10 (positioned at the end of the chapter) presents the financial-equation results: the variable revenue attributed to each of the four drivers, plus fixed revenue. The first three lines in each block presents the coefficients for the average doctoral segment (for which *S-index*=0), the segment with the largest *S-index*, and the segment with the smallest *S-index*. The next three lines present the data for the three nondoctoral segments: private liberal arts colleges, private comprehensive institutions, and public nondoctoral institutions. The *BachAll* and *R&D* coefficients have been multiplied by 1000 and the *Bach* and *Master's* coefficients by 10 to improve readability. Blanks indicate the main effect did not enter the equation for the given field, and equality across the doctoral segments indicates that *S-index* and *toc* did not enter. The fields are arrayed in the following order: first the sciences, then math and the social sciences, and then engineering and computer science.

Table 9.11 summarizes the utility and endogenous-variable results for the doctoral segments (the only ones where these effects are relevant). The utility presentation is couched in terms of ratios, since the absolute values of the utilities are indeterminate. The first block shows the ratio of doctoral to faculty utility; the second, the ratio of PostDoc to faculty utility; and the third, the ratio of PostDoc to doctoral utility. The blanks indicate fields for which the lowest segment's computed utility is negative: we set the utility coefficient to zero in each such case, but we note that in no case was the amount of negative excursion large.

The exogenous-variable effects are presented in Tables 9.12 and 9.13 (at the end of the chapter): the first summarizes the Faculty equation; the second, the Doctoral and Postdoctoral equations.

Conclusion

This chapter describes our model for the effects of degree production and sponsored research funding on doctoral and postdoctoral enrollment and faculty size. In Section I, we formulated the structural equations and developed appropriate estimators. Section II uses the algorithms to generate parameter estimates and simulate departmental resource allocation behavior for physics. Section II extends the physics parameter estimations to all 12 fields included in our science and engineering dataset.

Both the statistical and simulation results are satisfactory. The estimated equations fit the data well; nearly all the coefficients have the expected sign, and most of them are statistically significant. The estimates are generally stable as variables are added to and subtracted from the equations, and the estimates tend to show consistent patterns across fields. While the simulation algorithm had to be modified to enforce a two-sided financial constraint, the amended version produces credible simulation results. Hence we conclude that the model is sufficiently robust to use in our simulation of the supply and demand for science and engineering doctorates.

Endnotes

[1]We have data on the number of faculty in 1980 for doctoral-granting institutions, but such data do not exist for the nondoctoral-granting departments. For them, we use total science and engineering personnel as a surrogate for faculty. Since the nondoctoral institutions have little sponsored research, it is likely that most S&E personnel do in fact represent faculty.

Table 9.10 Model Estimates: Fixed and Variable Estimates

Segment		Biol	Chem	Physics	Geos	Math	Econ	Psyc	Chem. Eng.	Civil Eng.	Elec. Eng.	Mech. Eng.	Comp. Science
Bachall (000)	avg doc		123.6	184.2	104.2	218.5	203.6	127.9		92			67.3
	top doc		134.7	88.7	104.2	218.5	95.9	127.9		92			67.3
	low doc		132	235.6	104.2	218.5	223.6	127.9		92			67.3
	pr LAC	187.4	143.2	145.1	105.5	165.1		228.8		52.2			28.3
	pr comp	101	166	136.4	27.5	271.2	191.8	175.4	216		191.8	-50.7	63.5
	public	262.6	185.1	170.2	117	347.7	104.9	220.2	6.2	93.9	104.9	300.2	52.2
Bach (0)	avg doc	177.1	12.7	148.4	63.8	138.7	46.7	16.3	28.1		120	78.2	31.5
	top doc	150.8	36.7	208.7	275.9	150	4.1	16.3	21.6		120	47.1	31.5
	low doc	191.3		115.9	-9.5	155.9	54.7	16.3	21.6		120	47.1	31.5
	pr LAC	7.8	26	32.5	31.9	33.7	40.2	6.7	99.1	52.6	40.2	26	21.2
	pr comp	25.5	42.4	91.9	69.6	33.6	31.4	38.1	6.6	57.6	31.4	19.5	34.4
	public	22.2	66	42	40.8	31.8	41.5	37.6	33.6	65.1	41.5	44	28.1
Mast (0)	avg doc		109.8	1149.8				108.9	311.5	310.2		296.1	106
	top doc		11	287.9				444.9	311.5	368.4		460.2	329.9
	low doc							49.8	311.5	299.4		264.1	91.3
	pr LAC	47.2				46.9		26.4		15.2		10.6	
	pr comp							35	126.6				14.1
	public	115.7	48.4		0.8	48.7		18.6	132.7			110.9	11.5

R&D (000) avg doc	403.3	552	245.6	158	977	446.4	512.8	310.5	446.4	702	263.2	228.2
top doc	281.7	526.3	77.3	-78.6	977	284	512.8	263.7	96.7	325	-4.2	-13.8
low doc	468.6	574.1	254.1	182	977	259.6	512.8	263.7	511.4	673.4	315.4	244
pr LAC	435.1	418.2		25.3	1890.9		520.4				1436.4	
pr comp	675.6	691.5	288.8	193.1	233.7	1860.9	607.2	43.2		1860.9	475.3	289.9
public		370.4	680.7	62.7	860.1			198.2			325.8	217.6
Fixed (avg) top doc	97.9	-373	-118.3	-333.7	-52.3	-282.5	-824.3	-250.1	-155.3	-88.5	-112.4	-321.1
avg doc	-467.4	-351.3	1431.6	389.4	-214.2	1066.3	-495.7	-120	-183.7	901.1	-495.9	-224.9
low doc	401.2	-374.7	-380.6	-375.9	-40.8	-576.9	-822.4	-266.8	-209.7	-355.7	-132.9	-301.5
pr LAC	-11.9	-5	8.7	-11.1	-7.8	-39	1.2	-102.9	-33	-39	-19.2	-16.6
pr comp	-17.6	1.6	6.6	-11.8	22.2	-14.5	38.5	-22.2	0.5	-14.5	39	-4.5
public	-46.6	-24.2	-13.5	-1.9	8.4	-17.8	-18.8	32.4	-36.3	-17.8	-28.7	-11.9

Table 9.11   Model Estimates: Utility and Endogeneous Effects

	Segment	Biol	Chem.	Physics	Geos	Math	Econ.	Psyc	Chem. Eng.	Civil Eng.	Elec. Eng.	Mech. Eng.	Comp. Science
Utility Doc/Fac	avg doc	0.08	0.08	0.14	0.18	0.28	0.11	0.27	0.17	0.16	0.04	0.13	0.08
	top doc	0.08	0.08	1.17	1.73	0.85	0.92	1.57	0.93	1.08	0.26	0.7	0.99
	low doc	0.08	0.08	0.01	0.03	0.11		0.04	0.02			0.02	0.03
Utility PD/Fac	avg doc	0.99	0.2	0.05	0.51	0.5	0.18	0.11	0.37	0.26	0.31	0.31	0.16
	top doc	0.99	1.07	2.57	4.45	2.97	1.2	4.27	0.72	1.36	2.43	1.9	2.92
	low doc	0.99			0.11				0.31	0.05			
Utility PD/Doc	avg doc	12.99	2.34	0.37	2.77	1.82	1.64	0.4	2.21	1.61	8.25	2.38	1.93
	top doc	12.99	12.66	18.12	24.24	10.75	10.96	15.76	4.28	8.49	64.75	14.65	34.62
	low doc	12.99			0.6				1.81	0.33			
Endogen PD [Fac]	avg doc			0.303				0.082					
	top doc			0.303				0.082					0.013
	low doc			0.303				0.082					0.013
Endogen Fac [Doc]	avg doc	-0.01	0.081	0.184	0.166	0.093	0.061	0.032	0.001	0.003	-0.058	0.032	-0.006
	top doc	0.25	-0.045	-0.175	0.012	-0.168	-0.062	0.009	-0.102	-0.061	-0.153	-0.076	-0.042
	low doc	0.119	0.118	0.231	0.138	-0.168	0.084	0.036	0.021	0.015	-0.039	0.054	-0.004

Table 9.12   Model Estimates: Exogeneous Effects for Faculty

Segment		Biol.	Chem.	Physics	Geog. Sci.	Math	Econ.	Psyc.	Chem. Eng.	Civil Eng.	Elec. Eng.	Mech. Eng.	Comp. Science
Bachall (000)	avg doc		1.683	2.847		6.138	0.735	2.124		0.715	2.064		1.192
	top doc		1.683	2.847		6.138	0.735	2.124		0.715	2.064		1.192
	low doc		1.683	2.847		6.138	0.735	2.124		0.715	2.064		1.192
	pr LAC		9.427	8.354	6.222	10.763						0.386	2.919
	pr comp	8.067	10.717	7.639	1.796	19.973	8.093	10.793	6.484	1.59	8.093	12.952	6.321
	public	9.404	7.269	6.596	6.556	14.515	4.167	9.545	1.287	2.856	4.167	3.328	3.153
Bach (0)	avg doc	3.441	0.077	1.876	0.626	2.669	0.437	0.451	0.478	0.344	0.49	0.836	0.48
	top doc	3.441	0.077	1.876	0.626	2.669	0.437	0.451	0.478	0.344	0.49	0.836	0.48
	low doc	3.441	0.077	1.876	0.626	2.669	0.437	0.451	0.478	0.344	0.49	0.836	0.48
	pr LAC	1.132	1.311	2.073	1.299	1.755	1.387	1.232	1.256	1.66	1.387	0.931	1.2
	pr comp	1.863	2.145	3.454	2.841	1.076	1.09	2.426	0.901	2.274	1.09	0.479	1.377
	public	1.335	3.005	1.077	1.529	1.862	1.364	1.53	1.044	3.291	1.364	1.395	1.212
Mast (0)	avg doc	0.987		0.827			1.471	1.92	2.977	3.003	5.944	1.581	1.432
	top doc	0.987		1.323			1.333	1.92	2.977	3.156	5.944	4.655	1.432
	low doc	0.987		0.559			1.333	1.92	2.977	3.156	5.944	0.981	1.432
	pr LAC	1.39	3.854	7.585	1.427	2.689	2.516	1.336	2.279			0.549	0.543
	pr comp	2.637	1.229	12.601	0.774	1.471		0.372					
	public	6.71											
R&D (000)	avg doc	6.432	5.11	2.013	0.37	22.27	6.344	9.73	3.761	3.629	0.646	1.065	2.242
	top doc	6.432	5.11	2.013	0.37	22.27	6.344	-25.778	3.761	3.629	0.646	2.877	2.242
	low doc	6.432	5.11	2.013	0.37	22.27	6.344	15.978	3.761	3.629	0.646	0.06	2.242
	pr LAC	27.15	14.182	17.208		96.072		44.983				3.427	
	pr comp	26.384	27.009	11.466	12.79	5.923	72.907	9.638	5.854			19.904	3.256
	public	2.766	12.677	34.19	2.683	88.541	5.106	7.261	1.281			19.548	17.854

Table 9.13   Model Estimates: Exogeneous Effects for Doctoral and Post-Doctoral Students

Segment	Biol.	Chem.	Physics	Geo. Sci.	Math	Econ.	Psyc.	Chem. Eng.	Civil Eng.	Elec. Eng.	Mech. Eng.	Comp. Sci.
Doctoral Students												
Bachall avg doc	20.063	13.316	9.098	12.11	11.827	13.435	14.333		8.481			9.374
(000) top doc	64.665	32.573	34.569	12.11	29.031	13.435	14.333		8.481			9.374
low doc	-3.864	8.572	5.728	12.11	6.899	13.435	14.333		8.481			9.374
Bach avg doc	12.187		6.273	8.492	4.716	0.883		3.669	2.466	7.854	3.968	6.322
(0) top doc	4.273		6.273	8.492	4.716	0.883		1.748		7.854	6.112	6.322
low doc	16.433		6.273	8.492	4.716	0.883		4.048		7.854	3.55	6.322
R&D avg doc	11.934	29.182	11.627	4.207	54.846	31.457	30.002	26.117	41.395	64.194	30.317	
(000) top doc	10.883	21.199	4.178	4.207	54.846	17.233	142.28	10.287	23.359	21.557	1.732	
low doc	12.498	31.15	12.613	4.207	54.846	34.106	10.248	29.236	44.751	64.343	35.896	
Post Doctoral Students												
R&D avg doc	7.093	9.19	2.308	0.487	1.93	0.539	0.884	1.144	1.013	0.272	0.184	
top doc	7.71	10.306	1.438	0.141	3.3	1.491	0.884	4.663	0.721	0.194	0.184	
low doc	6.761	8.915	2.423	0.522	1.538	0.361	0.884	0.45	1.067	0.287	0.184	

References

Beightler, C. S., Phillips, D. T., & Wilde, D. J. (1979). *Foundations of optimization.* Englewood Cliffs, NJ: Prentice-Hall.

Bowen, H. R. (1980). *The costs of higher education: How much do colleges and universities spend per student and how much should they spend?* San Francisco, CA: Jossey-Bass.

College and University Personnel Association. (1982, June 2). Faculty salary survey. *The Chronicle of Higher Education.*

College and University Personnel Association. (1987, April 29). Faculty salary survey. *The Chronicle of Higher Education.*

College and University Personnel Association. (1993, March 31). Faculty salary survey. *The Chronicle of Higher Education.*

Hillier, F. S., & Lieberman, G. J. (1990). *Introduction to operations research* (fifth ed.). New York, NY: McGraw-Hill Publishing Company.

Hopkins, D. S. P., & Massy, W. F. (1981). *Planning models for colleges and universities.* Stanford, CA: Stanford University Press.

James, E. (1982). How nonprofits grow: A model. *Journal of Policy Analysis, 2,* 350–366.

James, E. (1990). Decision processes and priorities in higher education. In S. A. Hoenack & E. L. Collins (Eds.), *The economics of American universities,* 77–106. Albany, NY: SUNY Press.

10. Modeling Student Attainment of the PhD

Estimating the probability distribution for achieving the doctorate has proved to be a formidable challenge (see previous work by Bowen & Rudenstine, 1992). The problem stems partly from the complexity of the behavioral processes involved and partly from data limitations: frequency data are collected by graduation cohort rather than matriculation cohort, dropout data are not systematically collected, and matriculants data are considered too unreliable to be made available on an institution-by-institution basis.

We constructed the model presented in this chapter in order to incorporate probabilities for doctorate attainment and time to the PhD degree in the doctorate supply-demand simulation described in Chapter 3. The three sections of this chapter describe the model's structure and estimation procedure, present a detailed examination of doctoral attainment and time-to-degree for physics, and extend the physics results (in summary form) to the other 12 science and engineering fields included in our database. We shall refer to the structure as a "stochastic doctoral cohort model," for reasons that will become apparent.

Technical description

Data

We obtained the following data by discipline for some 210 US research and doctorate-granting institutions. The institution-level data were aggregated into 10 segments using the statistical methodology described in the appendix. The data and model structure apply to a single segment or to pooled data for all the segments.

g_{Tk} *Frequency distribution for time-to-degree,* from the National Research Council (NRC), custom tabulations of the Survey of Earned Doctorates for 5-year aggregations of graduation cohorts beginning in 1967 and ending in 1991 (i.e., 1967–71, 1972–76, 1977–81, 1982–86, and 1987–91), in total and broken down between US citizens and noncitizens. Each frequency distribution consists of 19 years-to-degree cells, plus a cell for ≥20 years and an "unknown" cell. (The last two cells were sparsely populated, so we felt justified in dropping them.) The dataset provided two versions of each frequency distribution:
- *Registered time-to-degree (RTD):* the time a student actually was registered in graduate school before obtaining the doctorate, for all doctorate recipients. Time between the baccalaureate and first graduate registration and "time off" during graduate study are excluded. These data seem to us to be the most relevant, given that we are going to use the model's results in the context of the academic production function.
- *Total time-to-degree (TTD):* the total time from bachelors to PhD degree, for all doctorate recipients. These data provide a useful contrast to the RTD data, as one can predict that the times to degree will be longer, the attrition rates higher, and the dispersion of graduation and dropout behavior greater.

n_t *Total fall enrollment of all graduate students (student census),* from CASPAR, annually from 1972 to 1990, in total and broken down between US citizens and others. We use the complete count of graduate students because we cannot distinguish between graduate students who intend to pursue a PhD from those with other intentions.

We could not obtain data on the number of dropouts or on the number of new students ("matriculants") by segment or citizenship. We do, however, have a limited time series consisting of aggregate matriculants data:

M_t annual data on the aggregate number of new students by discipline for the years 1982–90.

The number of matriculants for earlier time periods, for segments, and by citizenship will be approximated as described later.

Goal

The goal is to obtain consistent and efficient estimates for expected time to the doctoral degree and associated quantities, using the data described above. In particular, we want to estimate the transition probabilities needed by our simulation model:

$\Pr[G \in [(t, t+\Delta t) | S@t]$: the probability that a student will graduate in the time interval $(t, t+\Delta t)$, given that he or she has been a student for t periods;

$\Pr[D \in [(t, t+\Delta t) | S@t]$: the probability that a student will drop out in the time interval $(t, t+\Delta t)$, given that he or she has been a student for t periods.

Given the above, the probability that a student will stay in the program is

$$1 - \Pr[G \in (t, t+\Delta t) | S@t] - \Pr[D \in (t, t+\Delta t) | S@t].$$

We believe the structure may vary by citizenship, segment, and discipline, so we wish to produce estimates on a disaggregative basis.

Model

We postulate that each entering student can be characterized by two *transition propensity functions,* each depending on a student-specific parameter γ:

> $g[t \mid S@t,\gamma_g]$: propensity to graduate t periods after entering the program, given that he or she has remained a student to that point and behaves according to parameter γ_g;
>
> $d[t \mid S@t,\gamma_d]$: propensity to drop out t periods after entering the program, given that he or she has remained a student to that point and behaves according to parameter γ_d.

The propensity functions are probability densities rather than probabilities. Our assumed structure is independent of the student's year of entry, which implies that the parameters are identical for all student cohorts. (Estimating the model for subintervals of time allows different parameter values, however.) The parameters γ_g and γ_d are assumed to vary over the student population as described later.

Experience suggests that the graduation propensity evolves through a series of phases, where a student can be said to:

> gestate: a period of course-taking and other preparation, where the graduation propensity remains essentially at zero;
>
> progress: a period of intense research on one's dissertation, where the graduation propensity rises over time and eventually reaches a maximum value characterized by γ_g;
>
> sustain: a period where active research is mingled with other activities (including, for some, a full-time job), during which time the graduation propensity remains essentially constant;
>
> erode: a period where the impetus for research is under-

mined and discouragement may set in, during which time the graduation propensity declines;

stagnate: a period where the student remains in the program but languishes at the lower graduation propensity reached at the end of the erosion phase. (This phase is labeled "hang" in Figure 10.1.)

While similar phases could be identified for the dropout propensity, data limitations led us to assume that $d[t \mid S@t,\gamma_d]$ is time-invariant. If data on dropouts were to become available, the model could be extended to make the dropout propensity vary according to the phases given above or a different set of phases. The present dataset does not permit identification of the extended-model parameters, however.

The following structure maps these postulates into mathematical functions with the properties needed for parameter estimation: continuous first-order derivatives with respect to the parameters and integrability with respect to time.

$$
\begin{aligned}
g[t \mid S@t,\gamma_g] &= 0, & 0 \le t < t_P & \quad \text{(gestate)} \\
&= \frac{\gamma_g}{2}\left(1 - \cos\left[\frac{\pi(t - t_P)}{(t_E - t_P)}\right]\right) & t_P \le t < t_S & \quad \text{(progress)} \\
&= \gamma_g, & t_S \le t < t_E & \quad \text{(sustain)} \\
&= \frac{(1 + r)}{2} + \frac{(1 - r)}{2}\cos\left[\frac{\pi(t - t_E)}{(t_M - t_E)}\right] & t_E \le t < t_M & \quad \text{(erode)} \\
&= \gamma_g r, & t > t_M & \quad \text{(stagnate)} \\
d[t \mid S@t,\gamma_d] &= \gamma_d, & t \ge 0 & \quad (10.1)
\end{aligned}
$$

with additional parameters

t_P = beginning of the "progress" phase,

t_S = beginning of the "sustain" phase,

t_E = beginning of the "erode" phase,

t_M = beginning of the "stagnate" phase, and

r = ratio of the graduation propensity during the "stagnate" phase to that during the "sustain" phase.

There is a logical case for allowing r to vary among students, just like γ_g and γ_d. The model's mathematical structure would permit this, but the low incidence of graduation events at large t-values makes such a refinement impractical from the standpoint of identification.

Figure 10.1 interprets the parameters graphically by providing sample time shapes for the graduation and dropout propensities. It is apparent that the t-values are subject to the definitional constraints $0 \leq t_P < t_S < t_E < t_M$, which have been incorporated into the model. We also included the additional constraints $t_M \leq 20$ and $0 \leq r \leq 1$ to guard against instabilities stemming from weak identification.

Our final postulates concern the variation of γ_g and γ_d over the student population. These parameters are not observable, so the model must be couched in terms of their frequency distributions (Massy, Montgomery, & Morrison, 1974). We assume that γ_g and γ_d are distributed as independent gamma densities. The gamma density restricts the parameters to the positive real line, as required by the theory, and the result is tractable mathematically. We denote the parameters of the two gamma distributions as $\{\alpha_g, \beta_g\}$ and $\{\alpha_d, \beta_d\}$, where $E[\gamma] = \alpha \beta$ and $var[\gamma] = \alpha \beta^2$.

To trace the model's time evolution, we solve the following differential equation:

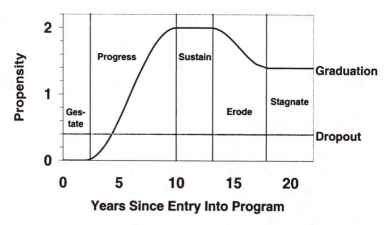

Figure 10.1   *Sample Time Shape for An Individual's Graduation and Dropout Propensities*

$$\frac{dh}{dt} = -(g[t \,|\, S@t,\gamma_g] + d[t \,|\, S@t,\gamma_d])h$$

$$\int \frac{dh}{h} = -\int (g[t \,|\, S@t,\gamma_g] + d[t \,|\, S@t,\gamma_d])dt \qquad (10.2)$$

$$\Rightarrow h[t \,|\, \gamma_g,\gamma_d] = h_0 e^{-(G[t \,|\, S@t,\gamma_g] + D[t \,|\, S@t,\gamma_d])}$$

where $h_0 = 1$ is the initial condition and $G[\cdot]$ and $D[\cdot]$ are the cumulative conditional propensities for graduation and dropout. The new quantity $h[t \,|\, \gamma_g,\gamma_d]$ represents the probability of remaining in the program t periods after entering, given that one is behaving according to γ_g and γ_d. That is:

$$h[t \,|\, \gamma_g,\gamma_d] = \Pr[S@t \,|\, \gamma_g,\gamma_d] \qquad (10.3)$$

The right-hand side of (10.2) is obtained by separately integrating the parts of (10.1) with respect to time, and then forming the definite integral for the whole time-line by taking the terminal value from each stage as the initial condition for the next one.

The probability density for graduating at t, conditional on γ_g and γ_d, can be written as:

$$g[t \,|\, \gamma_g,\gamma_d] = g[t \,|\, S@t,\gamma_g]h[t \,|\, \gamma_g,\gamma_d] \qquad (10.4)$$

Next we integrate out the unobservable γ_g and γ_d to get the unconditional densities for graduating, dropping out, and remaining a student at t, as illustrated here for graduation:

$$g[t] = \int_0^\infty \int_0^\infty g[t \,|\, \gamma_g,\gamma_d] \, f[\gamma_g]f[\gamma_d]d\gamma_g d\gamma_d$$

This reduces to the following general form:

$$g[t] = k_{g0}[t] \left\{ \int_0^\infty \gamma_g e^{k_{g,1}[t] + k_{g,2}[t]\gamma_g} f[\gamma_g]d\gamma_g \right\} \left\{ \int_0^\infty e^{k_{g,3}[t]\gamma_d} f[\gamma_d]d\gamma_d \right\} \qquad (10.5)$$

which integrates to a closed form when, as assumed, the $f[\cdot]$ are gamma densities. The constants k are functions of t and the parameters, and $d[t]$ and $\Pr[S@t]$ are obtained in a similar fashion.

The unconditional probabilities of graduating and dropping out in the interval beginning at t, which we denote by $\Pr[G{\in}(t,t{+}\Delta t)]$ and $\Pr[D{\in}(t,t{+}\Delta t)]$, are obtained from the definite integral of (10.5) and its $d[t]$ analog over the interval $(t,t{+}\Delta t)$. Then the conditional population probabilities for graduation and dropout—the transition probabilities which represent the goal of our analysis—are obtained by applying Bayes' Theorem:

$$\Pr[G{\in}(t,t{+}\Delta t)\,|\,S@t] = \frac{\Pr[G{\in}(t,t{+}\Delta t)]}{\Pr[S@t]}$$

$$\Pr[D{\in}(t,t{+}\Delta t)\,|\,S@t] = \frac{\Pr[D{\in}(t,t{+}\Delta t)]}{\Pr[S@t]} \qquad (10.6)$$

The time-integrations needed to obtain closed-form expressions for the unconditional graduation and dropout probabilities elude us, but numerical approximations can be obtained easily.

In their discussion of PhD completion rates and time-to-degree, Bowen and Rudenstine (1992, p. 106) describe a number of interpretive measures that can be calculated from our model. Their "truncated completion rate," the number of students of an entering cohort who have earned the doctorate within t years of graduate study, is equivalent to our $\Pr[G{\in}(0,t)]$ when estimated from "total time-to-degree" data. Their "minimum completion rate" is equivalent to the same quantity when one uses registered time-to-degree data for students doing all their work at the same institution. The asymptotic cumulative completion rate, the fraction of program entrants who eventually graduate, is simply $\Pr[G{\in}(0,t)]$ as t goes to infinity. Finally, we can calculate their median time-to-degree by solving for t in:

$$\Pr[G{\in}(0,t)] = 0.5\,\Pr[G{\in}(0,\infty)], \qquad (10.7)$$

which is the time-point where the cumulative completion rate reaches 50% of its asymptotic value. Notice that our completion rates and times to degree take account of population heterogeneity in the graduation and dropout propensities.

Bowen and Rudenstine (1992, p. 122) suggest that time-to-degree can be approximated by combining a normal distribution for "students who finish expeditiously" with "a series of three or four delays that impede the progress of other students." Our model presents a different approach though the population-heterogeneity idea carries over. We do not try to identify specific "delay points," (such as admission to candidacy or thesis-proposal approval), since the needed data are not available; however, the heterogeneity in the graduation and dropout propensities and the several time-stages built into the model produce similar results. Our model avoids the bias problem Bowen and Rudenstine (1992, p. 116) describe. They point out that a strong downward trend in matriculations will artificially inflate the time-to-degree statistics, because a disproportionate number of later-year graduates will have matriculated in earlier periods and taken a longer time to finish. Our model avoids this problem by modeling the population heterogeneity explicitly, so that each year's degree cohort contains the proper mix of fast and slow finishers regardless of the trend in matriculations.

Estimation

We estimate parameters by numerically minimizing the weighted sum of squared deviations between the data and the model's predictions. The predictions for time-to-degree frequency and student census are obtained from $Pr[G \in (t, t+\Delta t)]$, $Pr[S@t]$, and the estimated number of matriculants for each time period. Let $m[t]$ be a function (defined below) for calculating the matriculants in year t. Then the frequency and census predictions are:

$$g[T,k] = m[T - k]Pr[G \in (k, k + \Delta k)]$$

$$n[t] = \sum_{k=0}^{k^*} m[t - k]Pr[S@k] \qquad (10.8)$$

where $k^*=25$ approximates infinity for calculation purpose.

Obtaining $Pr[G \in (k, k+\Delta k)]$ by numerical integration using Mathematica's algorithm is too time-consuming for parameter estimation, since the function must be evaluated thousands of times. Hence we used the approximation:

$$Pr[G \in (t, t+\Delta t) | \gamma_g, \gamma_d] \approx \{G[t+\Delta t | S@t, \gamma_g] - G[t | S@t, \gamma_g]\}h[t | \gamma_g, \gamma_d] \quad (10.9)$$

where the gammas can be integrated out as shown in (10.5) for the density functions. We then evaluate $\Pr[G \in (k, k+\Delta k)]$ by summing small values of t over the interval $(k, k+\Delta k)$. Experimentation indicated that, in the worst case, 10 increments are sufficient to approximate the probability of graduating in a one-year interval to an accuracy of about 0.003, whereas the a single-increment error is 19 times larger. Hence we designed the estimator to use increments of between one and 10, depending upon the error sensitivity for the time-value in question.

As stated at the outset, we have aggregate matriculants time series only for the years 1982–90, whereas the estimator requires matriculants estimates for k^* periods prior to the first graduation cohort (1967). Therefore, we added an auxiliary model for projecting matriculants. We chose a cubic equation as the best compromise between flexibility and parsimony in the number of parameters:

$$m[t] = m[t_0]\{b_1(t - t_0) + b_2(t - t_0)^2 + b_3(t - t_0)^3\} \quad (10.10)$$

where t_0=1986 and b_1, b_2, and b_3 are estimated along with the model's other parameters. Setting t_0 at the midpoint of the matriculants data series allows us to interpret $m[t_0]$ as the average of the nine matriculants values. Hence $m[t]$ equals the average when $t = t_0$, and deviates smoothly from the average as t varies. We know the grand-total $m[t_0]$ for all the segments and citizenship classes, so we can estimate the fully pooled parameters without further difficulty.

The estimator minimizes the chi-square formed by combining the weighted sums of squared deviations between the actual and predicted values for the graduation-cohort frequencies, the student censuses, and the matriculants.

$$\chi^2 = w_1 \sum_T \sum_k \frac{(g_{Tk} - g[T,k])^2}{g[T,k]} + w_2 \sum_t \frac{(n_t - n[t])^2}{n[t]}$$
$$+ w_3 \sum_t \frac{(m_t - m[t])^2}{m[t]} \quad (10.11)$$

The weights should approximately equalize the three terms in accordance with generalized least squares practice. The weights are set twice: initially at one over the number of terms in the sum, and again, after the coarse-grain minimization using the pattern search procedure introduced below, to equalize the variances. The latter process solves three simultaneous linear equations which equate the three objective-function terms and constrains the sum of the weights to one.

The minimum chi-square procedure is consistent and asymptotically efficient (Rao, 1973, p. 352 and also Lehmann, 1991, pp. 446–48). The error variance equals χ^2/df, and the parameters' asymptotic variance-covariance matrix equals the error variance times one-half the inverse of the matrix of second order cross-partial derivatives of (10.11) with respect to the parameters. The requirement that the first derivatives of the propensity function's time parameters be continuous, stated just before (10.1), ensures that the second derivatives of (10.11) will be continuous as well, since (10.11) is based on the propensity-functions' time integrals.

The probability expressions comprising (10.11) are highly nonlinear in the parameters, so we minimize the chi-square by numerical means. Thirteen parameters must be accommodated, some of which are subject to constraints. After experimentation, we settled on the Hook-Jeeves "pattern search" procedure (Beightler, Philips, & Wilde 1979, p. 236 and also Schwefel, 1977, p. 43) for coarse minimization and Mathematica's "FindMinimum" function (Wolfram, 1991, p. 794) for fine-grained optimization after the pattern-search converged. The constraints were handled by penalizing the objective function according to $1000 (1+x)^{100}$, where x is the amount by which the constrain is violated. Both the pattern search and Mathematica procedures handle this formulation with no difficulty, and tests showed that the process converges to a single point regardless of the starting values.

While the methods just described are sufficient for estimating the model's parameters from fully-pooled data, additional procedures are needed to differentiate between US citizens and noncitizens and to disaggregate by segment. As stated earlier, the problem arises because we do not have matriculants data by citizenship category or segment. The problem is the same for both disaggregations, but since there are only two citizenship categories as opposed to 10 segments, we use different strategies.

To disaggregate by citizenship, we simultaneously estimate separate parameter vectors for the two categories together with a new parameter, the proportion of $m[t_0]$ who are US citizens. We designed the estimation program to constrain certain parameters to be equal for the two categories to save computation time, but we can run the estimator without these constraints if we wish.

To disaggregate by segment, we developed a procedure for allocating the aggregate $m[t_0]$ among the segments using information obtained from the model itself. First we estimate parameters for all segments combined, using pooled-segment results obtained as described in the previous paragraph. Then we reverse the chi-square minimization, so that the main parameters are fixed and the $m_s[t_0]$, are treated as unknowns. The final step minimizes the error sum of squares for the segments' graduation frequencies and student censuses with respect to $m_s[t_0]$, for all s, subject to the constraint that the sum of the $m_s[t_0]$ equals the known aggregate $m[t_0]$. The $m_s[t_0]$ represent the sum of matriculants for US citizens and noncitizens; the proportion of citizens for each segment will be determined during the segment-level parameter estimation.

Results for physics

Figure 10.2 presents graphical representations of the model as fit to fully-pooled registered time-to-degree data for physics and astronomy, and the underlying parameter estimates are presented in first column of Table 10.1. The model fits well, as can be seen from the Figure's upper panels. The first panel superimposes the model's predictions for graduation-frequency on the normalized data graduation-frequency points. The second does the same for the actual and predicted census figures, and the third shows actual and predicted matriculants. Notice that the predicted matriculants curve goes through the average matriculants value, as it must, but lies below the weight of the data-points for small matriculants values—this is required in order to optimize the fit of the census and frequency data, which also key off the matriculants parameters.

The left-hand panel in Figure 10.2's middle tier graphs the upward progress and eventual erosion of graduation propensity, as depicted by the solid curve, and dropout propensity, as depicted by the dashed straight line. The graphs are based on a hypothetical student with gamma-values at the medians of the two heterogeneity distributions. Apparently, propensity peaks at around 11 years, the

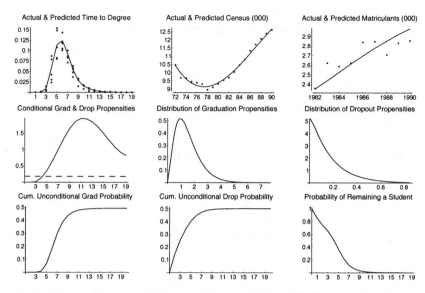

Figure 10.2   Registered Time-to-degree 12 Variable Estimation Physics and Astronomy: All Segments Combined 1967–1990

exact value being given by the "begin-sustain" and "begin-erode" parameters in the first column of Table 10.1. "Erosion" proceeds to a level, reached just after year 19, equal to 40% of the peak. The gestation period, represented by the "begin-progress" coefficient, is slightly longer than two years.

The standard errors of the time coefficient estimates are quite small—so small that one must look for a reason. The values for "begin-sustain" and "begin-erode" are of the order of 10^{-6}, which is an artifact of the constraint that prevents "begin-erode" from coming out less than "begin-sustain." The estimates are pinned against the constraint, and the strong curvature of the objective function drives down their standard errors. We suspect that a price must be paid in terms of higher-order effects, however, which may account for the relatively large standard errors of the propensities coefficients. We see, in fact, that none of the propensities coefficients are statistically significant, although the "graduate-shape" coefficient approaches significance at the 90% level.

Given these results, it seemed natural to disable the "erode" phase of the model by predetermining "erode-floor" at 1.0, which

Table 10.1   Parameter Estimates: Physics and Astronomy,
All Segments Pooled, 1967–1991

	Registered Time to Degree (RTD)		Total Time to Degree (TTD)
	12 vars	9 vars	9 vars
Time factors			
begin-progress	2.057	2.051	1.464
	(0.120)	(0.101)	(0.287)
begin-sustain	10.789	11.418	10.905
	(0.000)	(1.575)	(1.475)
begin-erode	10.789	18.000	18.000
	(0.000)	(na)	(na)
begin-hang	19.380	19.000	19.000
	(0.884)	(na)	(na)
Propensities			
erode-floor	0.403	1.000	1.000
	(1.786)	(na)	(na)
graduate-shape	2.640	2.576	4.826
	(1.373)	(0.401)	(0.615)
graduate-scale	0.573	0.661	0.148
	(0.414)	(0.247)	(0.039)
dropout-shape	1.044	1.601	0.654
	(2.262)	(0.255)	(0.035)
dropout-scale	0.163	0.097	0.375
	(0.451)	(0.018)	(0.029)
Matriculants			
Linear	83.111	77.903	80.020
	(0.313)	(7.871)	(5.206)
Quadratic (x10)	−1.874	−2.707	−2.936
	(0.515)	(1.202)	(0.773)
cubic (x1000)	−0.274	−0.309	−0.307
	(0.027)	(0.046)	(0.029)
Sum of squared errors	151.5	148.0	136.3
Degrees of freedom	111	114	114
R-squared	91.4%	91.6%	93.0%

Note: the figures in parentheses are the standard errors of the coefficients; "0.000" indicates that the estimate was imaginary.

makes the values of "begin-erode" and "begin-hang" irrelevant. (We set the last two parameters at 18 and 19; however, their values do not affect the model's predictions.) The 9-variable estimates are shown in the second column of Table 10.2. The gestation period is virtually unchanged, and the sustain period begins only slightly later. The standard error of "begin-sustain" now is a reasonable 1.575; both it and the "begin-progress" coefficient are highly significant. The last line of the table shows no loss in explanatory power. The 9-variable model actually appears to fit slightly better, but the small difference is due to variations in the objective function weights and the precision of the numerical minimization.

Moving on down the columns of Table 10.1, we see that the graduation propensity coefficients do not change much between the 12-variable and 9-variable runs, and that the latter are significant statistically. The second panel in Figure 10.2's middle tier graphs the gamma distribution produced by the 12-variable coefficients, and the 9-variable distribution is virtually identical. The

Table 10.2   Summary Statistics: Physics and Astronomy, All Segments, 1967–1991

	RTD 12 vars	RTD 9 vars	TTD 9 vars
Graduation probability	49.4%	48.6%	44.8%
Exp. time to degree:			
10% of cohort	3.78	3.77	3.77
25% of cohort	4.54	4.53	4.80
50% of cohort	5.55	5.53	6.16
75% of cohort	6.77	6.75	7.79
90% of cohort	8.17	8.13	9.62
Graduation propensity:			
median	1.33	1.49	0.66
mean	1.51	1.70	0.71
coef. of variation	0.62	0.62	0.46
Dropout propensity:			
median	0.12	0.12	0.14
mean	0.17	0.15	0.25
coef. of variation	2.91	2.62	2.11
R-squared	91.4%	91.6%	93.0%

dropout coefficients do change: "shape" grows materially while "scale" shrinks. The 12-variable dropout propensity distribution, shown to the right of the one for graduation propensities, approaches an exponential. The 9-variable distribution (not shown), however, rises to a peak at about 0.05 before falling away. Finally, the matriculants coefficients do not vary materially between the 12-variable and 9-variable runs.

The three panels in Figure 10.2's bottom tier depict the cumulative unconditional graduation and dropout probabilities and the unconditional probability of remaining a student. The graduation probability becomes asymptotic to 0.494 and the dropout probability to 0.506, by which point the probability of remaining a student has gone to zero. The figures imply that just under 50% of entering students will eventually graduate. The 9-variable curves look almost exactly the same.

We estimated both models for total-time-to-degree (TTD) data, but show only the 9-variable result. For the 12-variable case, "erode-floor" was estimated at 1.0, which makes "begin-erode" and "begin-hang" impossible to identify. The standard errors of these coefficients were complex, which is an indication of degeneracy. The other 12-variable coefficients were virtually identical to the 9-variable ones shown in the table. The "begin-progress" and "begin-sustain" coefficients are smaller than their RTD counterparts, a surprising result but, for the latter coefficient, one that does not have statistical significance. The differences between the matriculants coefficients also lack significance.

The median time-to-degree for the TTD run, about 6.2 years, is shorter than the ones reported by Tuckman, Coyle, and Bae for 1967 and 1986: 6.34 years and 7.07 years, respectively (Tuckman, Coyle, & Bae, 1990, Table A1, Physics/Astronomy, p. 115). Our figure should represent something of an average of theirs: 6.7 years versus our 6.2 years. We suspect the difference stems from the rise of foreign doctoral students, which as we shall see in Table 10.6 have a shorter time-to-degree, and the censoring associated with using raw time-to-degree statistics. It would appear that our model picks up the foreign shift more quickly than the raw statistics do. Our time-to-degree results are also smaller than the figures for time from the bachelors to the doctorate, 6.5 years for 1960–1969, rising to 6.9 years for the 1970s, and 7.3 years for the 1980s (NSF, 1993); however, one would expect time from first graduate registration (our figures) to be smaller than the time from the baccalaureate.

The propensities coefficients differ by amounts that are significant both statistically and materially in terms of interpretation. To understand the differences, we examine certain summary statistics computed from the model and presented in Table 10.2. The first row reports the fraction of entering students who are expected to graduate—that is, the asymptote of the unconditional cumulative graduation curve shown in Figure 10.2. As indicated earlier, we estimate that 49.4% of entering students will attain the doctorate eventually. The next five figures give the time required for the cumulative graduation probability to reach the indicated percentage of its asymptotic value. For example, 10% of graduates will get their degrees in 3.78 years, 90% will have done so in 8.17 years, and the median time-to-degree is 10.55 years. The median, mean, and coefficient of variation (mean/standard deviation) of the graduation propensity distribution are given next, followed by the same quantities for the dropout propensity distribution. The value of R-squared is again provided at the bottom of the table.

We now see that, as expected, the expected time-to-degree is longer for TTD than for RTD. TTD's shorter "begin-sustain" interval is swamped by its substantially smaller graduation propensity values, which are called out by the mean and median of the propensity distribution. These, together with the slightly larger figures for dropout propensity, also account for the smaller asymptotic graduation probability.

Bowen and Rudenstine report the (asymptotic) cumulative probability of gaining a physics doctorate at 71%, based on "elapsed time-to-degree" (ETD) data for eight research universities between 1967 and 1971. The eight universities are: Berkeley, Chicago, Columbia, Cornell, Harvard, Princeton, Stanford, and UNC (Bowen & Rudenstine, 1992, p. 134). ETD represents the time between first graduate registration and degree, including "time off" during graduate study. Their 71% is considerably larger than our 45% to 50%, based on RTD and TTD, which should bracket ETD. Doubtless much of the difference is due to their much narrower institutional sample and shorter time period. Indeed, the cumulative completion rate for physics and mathematics combined did drop from 66.9% to 64.6% between the 1967–71 period and the 1972–76 period, suggesting a downward trend. It seems possible, however, that Bowen and Rudenstine's figures suffer from the bias discussed earlier, which arises from the combination of censured data and nonstationary matriculations.

Turning now to the median time-to-degree, Bowen and Rudenstine report 6.4 years, for physics, based on ETD data for all universities between 1967 and 1971 (Bowen & Rudenstine, 1992, p. 132). We found the median to be about 10.5 for RTD and 6.2 for TTD, for all universities between 1967 and 1990 (Table 10.2). Our somewhat shorter times suggest that efforts to speed up time-to-degree may be bearing fruit.

Table 10.3 gives results for the allocation of matriculants to segments. To simplify the presentation in this chapter, we grouped the segments into two categories for each type of control: the two most elite segments (group A), and the remainder (group B). The table compares the average matriculation predictions for the segment groups with allocations based on the group's proportion of graduate enrollees. The largest percentage correction occurs in Segment A, where the model predicts 4.9% more matriculations than would have been obtained with proportionate allocation. The correction is at the expense of Segment D, but the percentage difference in this larger segment is modest. The figures are quite close for the other two segments.

Table 10.4 presents summary results based on parameter esti-

Table 10.3   Estimated and Proportional Matriculants and Number of Degrees, by Segment: RTD, Physics and Astronomy

Segment	Degrees	Estimated Matriculants	Proportional Matriculants.	Percent Difference
Private				
A	5,542	387	369	4.9%
B	6,124	506	506	–0.1%
Public				
C	4,009	413	413	0.1%
D	12,530	1,391	1,409	–1.3%

Table 10.4   Summary Results by Segment Groupings: RTD, Physics and Astronomy

	Private		Public	
	A	B	A	B
Graduation probability	58.3%	56.3%	53.3%	51.7%
Expected years to graduation				
10% of cohort	3.66	3.44	3.94	3.90
25% of cohort	4.30	4.00	4.72	4.71
50% of cohort	5.14	5.29	5.73	5.79
75% of cohort	6.13	6.58	6.96	7.11
90% of cohort	7.22	8.05	8.34	8.62
R-squared	82.6%	89.0%	89.2%	93.0%

mates for the four segments. It is immediately apparent that the asymptotic graduation probability declines with segment: from 58.3% for Segment A to 51.7% for Segment D. Moreover, the private sectors' probabilities dominate those in the public sector. Segment A's probability is about 10% greater than Segment C's (58.3/53.3), and the disparity between Segments B and D is about the same. Looking to the expected time to graduation, we see that students are more likely to graduate quickly in private-sector institutions than in public ones. Segment A also produces somewhat quicker graduations than Segment B, but the two public-sector segments exhibit little difference.

We note that the segment-level asymptotic graduation probabilities are all greater than the one for the pooled run (Table 10.2, RTD 9 years). The numbers of matriculants and the numbers of graduate students are conserved in the segment-level runs, so there has been no leakage of numbers out of the system. For the present we merely note this as an anomaly, which should be a subject for further study.

Table 10.5 presents summary statistics for four overlapping 10-year time intervals, covering the period from 1965 to 1990. Each interval contains two graduation cohort frequency distributions: for example, 1965–75 uses the 1967 and 1972 cohorts. All intervals use

Table 10.5   Results by 10-Year Period: RTD, Physics and
Astronomy; All Segments

	1965–75	1970–80	1975–85	1980–90
Graduation probability	53.7%	41.0%	50.4%	72.8%
Expected years to graduation				
10% of graduates	3.45	3.72	3.92	4.10
25% of graduates	4.24	4.46	4.69	4.88
50% of graduates	5.26	5.44	5.70	5.91
75% of graduates	6.47	6.64	6.94	7.18
90% of graduates	7.82	8.02	8.41	8.67
R-squared	91.6%	89.0%	90.1%	96.4%

the same matriculants' series (1982–90), though the earlier periods involve more backward extrapolation.

The results suggest that the asymptotic graduation probability and the expected time to the doctorate may have increased during the last 25 years. The graduation-probability increase occurred during the past decade, from the 40-50% range to more than 70%. (The 1970s seem to display a slight dip.) The time-to-degree has been creeping up steadily. For example, the median years to graduation increased from 5.26 in 1965–75 to 5.44 in 1970–80, and then to 5.70 in 1975–85 and 5.91 in 1980–90. The other percentiles show a similar pattern.

Table 10.6 presents pooled-segment results for US citizens and noncitizens. The comparison is quite striking. The asymptotic graduation probability for visa-holders is only two-thirds as large as that for citizens: 33% versus 53%. On the other hand, those visa-holders who do obtain their degrees will do so fairly quickly: The median time to graduation is only about 4.5 years compared to 5.5 years for their US colleagues. The timing distribution is sharply truncated for the noncitizens, with graduates requiring five years to get their degrees. The distribution for citizens shows the usual long tail, with 10% of graduates taking more than eight years. While we have not calculated the statistical significance of these deviations, the differences in the underlying parameters run to as much as several times

Table 10.6   US Citizens and Visa Holders: RTD, Physics and Astronomy, All Segments, 1965–1980

	US	Visa
Graduation probability	53.9%	33.2%
Expected years to graduation		
10% of graduates	3.74	3.59
25% of graduates	4.49	4.09
50% of graduates	5.49	4.46
75% of graduates	6.72	4.79
90% of graduates	8.13	4.74
R-squared	97.2%	

their standard errors. The value of R-squared, which applies to both citizenship categories, is a very respectable 97%.

Table 10.7 gives the conditional probabilities for graduation and dropout, which provided our original motivation for developing the model. Each figure represents the probability of a transition away from student status during the year, given that the person was a student at the beginning of the year. For example, new students have no chance of graduating during their first year, but 10.8% of the US citizens and 18.5% of visa-holders are likely to drop out. For each group, the probability of remaining a student is one minus the sum of the two transition probabilities.

The graduation probabilities rise steadily until the ninth or tenth year, after which they decline moderately. The dropout probabilities decline steadily; they become smaller than the graduation probabilities at around the fourth or fifth year. The probability of remaining a student reaches a minimum between the eighth and tenth year and then rises modestly. One should be careful when interpreting the outers' probabilities, however, since the data contain few observations for student tenures greater than eight or 10. Moreover, our assumption of a time-invariant dropout propensity subject to population heterogeneity guarantees that the population dropout transition probability will decline monotonically. As noted earlier, the time-invariance assumption arose as a matter of neces-

Table 10.7   Transition Probabilities: US Citizens and Visa Holders, RTD, Physics and Astronomy, All Segments, 1965–1980

Years in program	Graduation probability		Dropout probability		Continuation probability	
	US	Visa	US	Visa	US	Visa
1	0.0%	0.0%	10.8%	18.5%	89.2%	81.5%
2	0.0%	0.1%	10.4%	17.6%	89.6%	82.3%
3	1.4%	2.4%	10.0%	16.8%	88.6%	80.8%
4	9.4%	7.9%	9.3%	15.7%	81.3%	76.4%
5	21.6%	15.6%	8.5%	14.4%	69.9%	70.0%
6	32.3%	23.7%	7.6%	13.2%	60.1%	63.1%
7	38.7%	30.9%	7.0%	12.1%	54.3%	57.0%
8	41.2%	36.4%	6.7%	11.2%	52.1%	52.4%
9	41.1%	39.9%	6.5%	10.5%	52.4%	49.6%
10	39.6%	41.7%	6.4%	10.0%	54.0%	48.3%
11	37.4%	42.0%	6.3%	9.7%	56.3%	48.3%
12	34.9%	41.1%	6.3%	9.4%	58.8%	49.5%
13	32.3%	39.4%	6.2%	9.3%	61.5%	51.3%
14	29.7%	37.0%	6.2%	9.2%	64.1%	53.8%
15	27.2%	34.3%	6.1%	9.1%	66.7%	56.6%
16	24.8%	31.9%	6.1%	9.0%	69.1%	59.1%
17	22.6%	29.8%	6.1%	8.9%	71.3%	61.3%
18	20.4%	27.9%	6.0%	8.8%	73.6%	63.3%
19	18.4%	26.3%	6.0%	8.7%	75.6%	65.0%
20	16.7%	24.9%	5.9%	8.6%	77.4%	66.5%

sity given the absence of dropout data, not as a judgment about behavior.

Results for all fields

Having established the model's efficacy for physics, we proceeded to run it on data for the other fields included in our science and engineering dataset. The dataset consists of thirteen fields. Chapter 5 presents some summary information about the differences among fields. In the interest of space, we omit more detailed presentation of the parameter values for each field.

Conclusion

We have presented a stochastic model for predicting the asymptotic graduation probability, the expected time to the pH degree, and related quantities. The model was derived from assumptions about the time path of an individual's graduation and dropout propensity, and the distribution of these propensities over the population of entering students. By modeling population heterogeneity explicitly, the model avoids the data censoring problems that have been cited in the recent literature. Parameter estimates are obtained by a minimum chi-square procedure, which is consistent and asymptotically efficient. Parameters can be estimated for US citizens and noncitizens, for institutional segments, and for time intervals of fairly short duration. While requiring substantial computation effort, the parameter estimation procedure appears to be robust as assessed by the interpretability of results for various cuts at the data and by the standard errors of the coefficients.

References

Beightler, C. S., Phillips, D. T., & Wilde, D. J. (1979). *Foundations of optimization*. Englewood Cliffs, NJ: Prentice-Hall.

Bowen, W. G., & Rudenstine, N. L. (1992). *In pursuit of the PhD*. Princeton, NJ: Princeton University Press.

Lehmann, E. L. (1991). *Theory of point estimation*. Pacific Grove, CA: Wadsworth & Brooks/Cole Advanced Books and Software.

Massy, W. F., Montgomery, D. M., & Morrison, D. G. (1974). *Stochastic models of buying behavior*. Cambridge, MA: MIT Press.

National Science Foundation. (1993). *Science and engineering doctorates: 1960–91*. NSF 93-301. Detailed Statistical Tables. Washington, DC, 153–154.

Rao, C. R. (1973). *Linear statistical inference and its applications*. New York, NY: John Wiley.

Schwefel, H. P. (1977). *Numerical optimization of computer models*. New York, NY: John Wiley & Sons.

Tuckman, H., Coyle, S., & Bae, Y. (1990). *On time to the doctorate: A study of the increased time to complete doctorates in science and engineering.* Washington, DC: National Academy Press.

Wolfram, S. (1991). *Mathematica.* Redwood City, CA: Addison-Wesley.

11. Modeling Faculty Career Patterns

This chapter reports the work to estimate a model of faculty career patterns to support the PhD production and employment modeling project. In addition to its use in this project, the model holds promise of forecasting and budgeting at various levels of aggregation: university, state college system, or nationally.

Previous work on faculty careers has not analyzed transitions of individual faculty members over time. The most comprehensive survey of faculty is the National Survey of Postsecondary Faculty, which contained questionnaires for individual faculty and departments. The questionnaire for individual faculty members in 1987 asked faculty to identify their current rank, tenure status, and the year each was attained. From that, one can make some limited probability calculations, but the survey included only faculty currently employed in higher education institutions, not those who had changed jobs, retired, or died. As a result, the data do not account for all faculty career transitions. The NSOPF questionnaire for departments asked for the number of hires, departures, and tenure decisions during a single year (1986–87), but not by rank of faculty affected. From the departmental information, one year's transitions can be computed as an aggregate.

The work described in this paper uses data on individual faculty members in 13 science and engineering fields, across a wide spectrum of institutions. Each faculty member is tracked over the period 1968–1992 through hiring, promotions, and departure. With this information, we compute transition probabilities from rank to rank and for departures from each rank.

Using the data from different types of institutions over the 24-year period, we will test hypotheses about which factors related to institutions, disciplines, and time significantly affect the transition probabilities. The most interesting of the results concern the probability of promotion for an assistant professor, and the probability of departure for a full professor. These two transitions govern much of the flexibility that departments face in determining the size of their faculty under the tenure system.

Model

The model is described in Chapter 6. Figure 6.1 illustrates a Markov model of faculty careers from the perspective of an academic department. Careers cover the three rank stages: assistant professor, associate professor, and full professor. Faculty move from rank to rank or leave the department.

We estimate a multinomial logit model corresponding to Figure 6.1, with fixed effects for institutional type and academic field. (For a complete reference on the multinomial logit, see Amemiya, 1985, page 295.) Institutional control and type are defined by Carnegie classification: private research university, public research university, private doctoral-granting, public doctoral-granting, public comprehensive, and private liberal arts. We tested several specifications for time, including separate fixed effects for each year in the sample and a linear time trend.

This allows us to test hypotheses related to questions such as:

- Does institutional type, academic field, or year significantly affect the probability of promotion for an assistant professor?
- Is the probability of departure for full professors affected by institutional type, academic field, or year?

Data

We solicited faculty rosters from 10 colleges and universities, spanning the range from liberal arts colleges to major research universities. We attempted to obtain rosters for the fields listed in Table 6.1, at three-year intervals from 1968 to 1992 (i.e., 1968, 71, 74, 77, 80, 83, 86, 89, and 92). Except for one of the public research universities, we were able to obtain data for the entire time period. Table 11.1 shows the distribution of institutions by type of control and Carnegie classification.

From the rosters supplied, we traced individual faculty members at the three-year intervals listed above through the academic ranks: assistant, associate, and full professor. From the individual-level data, we computed the number of transitions along each arrow of Figure 6.1, for each three-year period, for each department in each school.

The aggregate transition frequencies (combining all fields, institutions, and three-year time periods) are reported in Table 11.2.

Table 11.1   Distribution of Schools

Carnegie type	Public	Private
RU	2	2
Doc	1	1
Comp	2	0
LAC	0	2

Table 11.2   Distribution of Transitions in the Data

			To			
		Out	Asst	Assoc	Full	Total
	Out		460	216	255	931
From	Asst	310	398	353	17	1078
	Assoc	176		468	369	1013
	Full	354			594	948
	Total	840	858	1037	1235	3970

The table shows the number of faculty moving "from" a given state (Out, Asst, Assoc, or Full), "to" a given state over a three-year period.

Because faculty members are observed every three years, it is possible for our data to show a transition from Assistant to Full Professor over a three-year interval, skipping the associate stage. Because the very few such observations did not allow for econometric precision, they are treated as transitions from assistant to associate in the estimations, with no material difference in the results. About 5% of all assistant professor promotions fell into this category.

Estimation and analysis

Estimations were performed using Stata 3.1 for three models: assistant, associate, and full. Tables 11.6 through 11.8 contain the results of the estimations, which are analyzed here. Despite testing several specifications for time effects, no significant effects were detected. Some of the tables report results from a model with a linear time trend in the underlying regressions, although the effects are insignificant. There may be time interaction effects with institutions, or specific fields, but this data set is not large enough to test for such effects.

The multiple equation models for assistant and associate professors are difficult to interpret directly, since changes in a parameter affect all the estimated equations. Therefore, this analysis relies on comparisons of the computed probabilities to illuminate the significance of the estimation results.

The results in Tables 11.3 and 11.4 present probabilities based on the estimated coefficients in Table 11.6. Using the estimated coefficients of the model, we calculated the fraction of transitions that represent promotions (to associate or full professor) as opposed to departures. A low percentage is consistent with tougher tenure standards. Figure 6.2 illustrates our results using coefficients for biology over the period 1989–92. Other fields and time periods display the same pattern.

For assistant professors, the results of the estimation indicated that institutional type and control were significant. The estimated

Table 11.3   Assistant Professors Estimated Promotions as Fraction of Transitions (Biology, 1968–71 and 1989–92)

	1968–71	1989–92
Private Research	47.0%	54.5%
Public Research	66.1%	72.5%
Private Doctoral-Granting	68.0%	74.1%
Public Doctoral-Granting	68.6%	74.7%
Public Comprehensives	65.1%	71.6%
Private Liberal Arts	48.8%	56.3%

Table 11.4   Assistant Professors Estimated Promotions as Fraction of Transitions (Private Research Universities, 1968–71 and 1989–92)

Field	1968–71	1989–92
math*	22.7%	28.4%
econ*	27.3%	33.7%
csci*	28.5%	35.0%
phys*	34.5%	41.6%
mech	35.1%	42.2%
psyc	39.8%	47.2%
chem	42.7%	50.2%
biol	47.0%	54.5%
civl	47.4%	54.9%
ceng	55.3%	62.6%
geos	56.2%	63.5%
elec	63.5%	70.2%
indu	68.7%	74.8%

*Denotes effects significant at the 5% level (asymptotic), from logit with biology as comparison field, with linear term for time.

effects are consistent with the hypothesis that there are two promotion probabilities operating. The lower probability of promotion is associated with

- Private research
- Private liberal arts

A higher probability of promotion is associated with

- Public research
- Private doctoral-granting
- Public doctoral-granting
- Public comprehensives

To explore difference across fields, Figure 6.3 presents the same percentages for every field in private research universities, 1989–92. Based on joint tests of equality of coefficients, differences in Figure 6.3 of about 20 percentage points or more are significant. For example, we cannot reject the hypothesis that math, economics, computer science, and physics are all governed by the same model. But this group of fields differs significantly from biology and chemistry.

For the full professor estimation, shown in Table 11.8, the coefficients can be interpreted directly. A positive coefficient indicates that the effect makes full professors more likely to remain and less likely to leave. Table 11.5 shows the computed probabilities for every field in Private Research Universities, 1989–92.

Table 11.5 Full Professors Estimated Transitions (Private Research Universities, 1989–92)

Field	Percent Remaining	Percent Leaving
indu	83.1%	16.9%
geos	84.5%	15.5%
biol	84.8%	15.2%
psyc	85.4%	14.6%
civl	85.6%	14.4%
math	87.9%	12.1%
chem	87.9%	12.1%
phys*	89.3%	10.7%
csci*	90.0%	10.0%
econ*	90.1%	9.9%
ceng	91.9%	8.1%
elec*	92.1%	7.9%
mech*	93.1%	6.9%

*Denotes effects significant at the 5% level (asymptotic), from Logit with biology as comparison field, with linear term for time.

Table 11.6 Assistant Professors Multinomial Logit Estimation

	Coefficient	Standard Error	z	P>|z|
Asst				
ceng	0.5628	0.4586	1.227	0.220
chem	0.1509	0.2209	0.683	0.495
civl	−0.0265	0.3198	−0.083	0.934
csci	−0.2308	0.2187	−1.055	0.291
econ	−0.3847	0.2123	−1.812	0.070
elec	0.4593	0.3464	1.326	0.185
geos	0.2028	0.3241	0.626	0.532
indu	0.8084	0.6731	1.201	0.230
math	−0.6927	0.1994	−3.475	0.001
mech	−0.5342	0.3783	−1.412	0.158
phys	−0.5213	0.2171	−2.401	0.016
psyc	−0.3290	0.2273	−1.447	0.148
priru	0.4909	0.1698	2.891	0.004
pubru	0.7749	0.1838	4.216	0.000
pridoc	1.4916	0.2582	5.777	0.000
pubdoc	0.7193	0.2111	3.408	0.001
comp	0.5377	0.1811	2.969	0.003
lac	0.3955	0.2723	1.453	0.146
Assoc				
ceng	0.3254	0.4767	0.683	0.495
chem	−0.1546	0.2318	−0.667	0.505
civl	0.0930	0.3251	0.286	0.775
csci	−0.7576	0.2363	−3.206	0.001
econ	−0.8445	0.2336	−3.615	0.000
elec	0.6555	0.3474	1.887	0.059
geos	0.3723	0.3280	1.135	0.256
indu	0.9704	0.7006	1.385	0.166
math	−1.1277	0.2202	−5.122	0.000
mech	−0.4715	0.3823	−1.233	0.217
phys	−0.5076	0.2237	−2.269	0.023
psyc	−0.3134	0.2340	−1.339	0.180
priru	0.0104	0.1802	0.058	0.954
pubru	0.8117	0.1874	4.332	0.000
pridoc	0.9706	0.2844	3.413	0.001
pubdoc	0.8839	0.2136	4.138	0.000
comp	0.7216	0.1821	3.962	0.000
lac	0.1387	0.2992	0.463	0.643

Note: The omitted field is biology. Number of observations = 2235. Log Likelihood = −2341.2677. Pseudo R^2 = 0.0465. Assistant professors can move to four states in the Markov chain: ASST, ASSOC, or OUT. The model is estimated with a two-equation multinomial logit, with OUT as the comparison state. This model does not include a time specification.

Table 11.7 Associate Professors Multinomial Logit Estimation

	Coefficient	Standard Error	z	P>|z|
Assoc				
ceng	0.8455	0.7628	1.108	0.268
chem	0.1783	0.3195	0.558	0.577
civl	−0.1041	0.3618	−0.288	0.774
csci	−0.2228	0.2918	−0.763	0.445
econ	−0.2962	0.2945	−1.006	0.315
elec	0.3308	0.3634	0.910	0.363
geos	0.2360	0.3900	0.605	0.545
indu	−1.2350	0.5890	−2.097	0.036
math	0.0668	0.2872	0.233	0.816
mech	0.3300	0.3894	0.847	0.397
phys	0.0984	0.3004	0.328	0.743
psyc	−0.1725	0.2710	−0.636	0.525
priru	1.5227	0.2352	6.475	0.000
pubru	1.6106	0.2161	7.453	0.000
pridoc	1.9154	0.3409	5.619	0.000
pubdoc	2.0943	0.3114	6.725	0.000
comp	1.6767	0.2202	7.615	0.000
lac	1.6347	0.4485	3.645	0.000
Full				
ceng	0.9110	0.7748	1.176	0.240
chem	0.2728	0.3289	0.829	0.407
civl	0.1772	0.3682	0.481	0.630
csci	−0.4027	0.3109	−1.295	0.195
econ	−0.1834	0.3055	−0.600	0.548
elec	0.3182	0.3753	0.848	0.397
geos	0.2962	0.3993	0.742	0.458
indu	−1.3448	0.6261	−2.148	0.032
math	−0.2538	0.3043	−0.834	0.404
mech	0.1113	0.4080	0.273	0.785
phys	0.0302	0.3114	0.097	0.923
psyc	−0.3498	0.2856	−1.225	0.221
priru	1.3448	0.2423	5.550	0.000
pubru	1.1992	0.2250	5.331	0.000
pridoc	1.2291	0.3627	3.388	0.001
pubdoc	2.0135	0.3165	6.362	0.000
comp	1.1086	0.2309	4.801	0.000
lac	0.8832	0.4874	1.812	0.070

Note: The omitted field is biology. Number of observations = 2261. Log Likelihood = −2083.1239. Pseudo R^2 = 0.1614. Associate professors can move to three states in the Markov chain: ASSOC, FULL, or OUT. The model is estimated with a two-equation multinomial logit, with OUT as the comparison state. This model does not include a time specification.

Table 11.8   Full Professors Logit Estimation

	Coefficient	Standard Error	z	P>|z|
ceng	0.5392	0.2925	1.843	0.065
chem	0.1574	0.1480	1.064	0.288
civl	−0.0644	0.1623	−0.397	0.692
csci	0.1149	0.1622	0.709	0.479
econ	0.3264	0.1613	2.023	0.043
elec	0.1727	0.1582	1.091	0.275
geos	0.0240	0.1790	0.134	0.893
indu	−0.2323	0.5039	−0.461	0.645
math	0.5191	0.1614	3.215	0.001
mech	0.2651	0.1918	1.382	0.167
phys	0.3255	0.1458	2.232	0.026
psyc	0.0419	0.1421	0.295	0.768
priru	1.8418	0.1181	15.601	0.000
pubru	1.4769	0.1124	13.141	0.000
pridoc	1.5344	0.2608	5.884	0.000
pubdoc	2.1775	0.1644	13.244	0.000
comp	1.7020	0.1216	13.992	0.000
lac	1.0508	0.2761	3.806	0.000

Note: The omitted field is biology. Number of observations = 6895. Log Likelihood = −2711.2252. Full professors can either remain full professors or leave. With only two states, the estimation is done with ordinary logit, giving coefficients for the probability of remaining (REMAIN). This model does not include a time specification.

Converting to annual transition probabilities

In order to use the models estimated here for an annual simulation of faculty transitions, one must convert the predicted transition probabilities, which represent three-year intervals. The calculation of this conversion is outlined below. Because our data are based on tracing a specific population of individuals over three-year intervals, we need not consider the continuing hiring of new members in the transition probabilities.

Consider a single state, with several possible transitions, for example, from ASST to ASST, ASSOC, or OUT. The estimation results in a predicted probability for each transition. For assistant professors, that is one transition remaining in the same state, ASST, and two transitions to other states. Let p_{OUT} represent the total annual probability of transitions out of the state to any other state and $1 - p_{OUT}$ the annual probability of remaining in the state. $\hat{p}_{OUT}$ is our calculated three-year probability of transitions out of the state. $\hat{p}_{OUT}$ satisfies the equation:

$$\hat{p}_{OUT} = p_{OUT}(1 + (1 - p_{OUT}) + (1 - p_{OUT})^2) \qquad (11.1)$$

This equation has a single real solution:

$$p_{OUT} = 1 - (1 - \hat{p}_{OUT})^{\frac{1}{3}} \qquad (11.2)$$

If p_1 is the probability of a transition out of the state, to a single new state, such as ASST to ASSOC, then the three-year probability, $\hat{p}_1$, must obey

$$\hat{p}_1 = p_1(1 + (1 - p_{OUT}) + (1 - p_{OUT})^2). \qquad (11.3)$$

Using the equations (11.1) and (11.3), we derive the relation for any transition probability,

$$p_1 = \hat{p}_1\left(\frac{p_{OUT}}{\hat{p}_{OUT}}\right). \qquad (11.4)$$

References

Amemiya, T. (1985). *Advanced econometrics*. Cambridge, MA: Harvard University Press.

12. Future Directions in Research

This research has broken new ground in several areas. But as in all research, there are lines of inquiry that we have not been able to explore fully and new lines of inquiry suggested by our results. This chapter sums up these opportunities, using the basic structure of the second half of the book as a template. In the following four sections, we discuss future opportunities for research in the overall labor market model, departmental decision-making, student attainment of the PhD, and faculty careers.

Overall labor market model

The big need, of course, is for qualitative research into the displacement of lesser degree holders by PhDs when there are positive base employment gaps. What is the profile of PhD jobs for new PhDs under these circumstances? To what extent does the PhD add value in different kinds of jobs as compared to lesser degree holders with good on-the-job training, and what are the relative costs of the two modes of preparation? Can a "breakeven" job type be discerned, and if so where does it stand in the profile? Such research would require extensive fieldwork rather than econometric modeling. If successful, the results would inform future quantitative research on the supply and demand for PhDs.

More work also remains to be done in the area of feedback and adjustments. We remain convinced that neither departments nor prospective doctoral students take close account of the doctorate employment gap, but there is a need to test this proposition more thoroughly than we have here.

Departmental decision-making

This model retains a few "loose ends" that deserve followup as part of some future project. Most important is to look the reasons that an enforced two-sided financial constraint is required for simulation. The model's built-in weighting scheme isn't working as we had expected, and the weights should be traced back through the estimator to make sure everything is consistent. Enforcing the financial constraint seems appropriate enough for simulation purposes, since departments surely should be treated as shunning deficits but spending all the money they can get. The approach, however, is not as theoretically attractive as the one-sided model.

The second task would be to replace the current piecewise-linear LP maximizing procedure with a proper quadratic programming algorithm. This would speed up computation, reduce complexity, and probably produce a better result. This is not a conceptually difficult task, but finding or developing a function that would run in or could be called from Mathematica proved to be beyond our capability on this project.

The third task involves further exploration of the parameter space for both the financial and production equations, especially for the fields other than physics where we had time for only a limited amount of testing. Once again, this is not conceptually difficult, but this task is sufficiently time-consuming to be beyond the hands-on capacity of the principal investigators. A more extensive search for possible rogue data points should take place prior to further empirical analysis.

Student attainment of the PhD

We offer several suggestions for followup work. First, the model as presently constituted estimates the matriculants' time trends separately for US citizens and visa holders, subject only to the constraint that the matriculants' fractions must sum to one in 1985. However, the equation probably can be modified so that the fractions sum to one in every year. (This requirement was added for purposes of simulation but not incorporated in the doctoral cohort estimation process.) Constraining all the matriculants fractions would produce a more precise and realistic model and also reduce the number of parameters. We would expect that the new model would produce results which are very similar to the ones presented here, but that, perhaps, the occasional anomalies would disappear.

A second followup task would be to find a better minimizing process for the model's sum of squared errors. The combination of pattern search and Mathematica's *FindMinimum* routine worked well enough for present purposes but without much slack. Most estimation runs took at least a day on the Unix workstations available in 1994–1995, and a few took as long as a week. Faster minimization (and advances in computational speed) would improve our ability to do sensitivity analyses and to troubleshoot difficulties. Alternatively, the model's objective function could be implemented in compiled C-code, or faster machines than the Sun workstations we used could be made available.

The combination of a revised model and faster parameter estimation might make it possible to run the model for individual segments rather than groups of segments. This would utilize the available data to greater advantage and provide more insight about intersegment differences. If individual-segment estimation proves unfeasible, one surely could break the "low-public" group, which for most fields now contains four segments, into two groups of two segments each. Additional analyses also could be done on shorter time periods, along the lines of the analysis reported for Physics in Table 5.5.

The fourth followup task involves processing the data for total time-to-degree. As noted earlier, we obtained data from the NRC for total time-to-degree as well as registered time-to-degree, but we have not had time to process the total time-to-degree data. While we believe the registered time-to-degree data are more relevant (certainly for the purposes of our simulation), the total time-to-degree data may produce additional insights. We believe, however, that tasks one through three, above, should be accomplished before turning to the total time-to-degree data.

In conclusion, we suggest that the model, as revised, be used on a regular basis to monitor trends in PhD production in the United States and perhaps in other countries where the requisite data exist. The new data that become available every year are too voluminous for easy interpretation, and we believe that the doctoral attainment model reported here could improve the situation materially. Additional work on institutional segmentation would be needed, since our segments were designed for a purpose other than monitoring PhD production. Indeed, it might even prove feasible to run the model on data for individual institutions. How the model might be used for ongoing monitoring purposes should be a subject for further study.

Faculty careers

The faculty career data we used were limited to a small set of institutions because of the lack of a national data source on faculty career patterns. Until such a data source is created and maintained, we will not be able to make national estimates of the rates of hiring, promotion, and departure. We hope that our early work in this vein will stimulate interest in collecting more comprehensive data on a regular basis, through the National Survey of Postsecondary Faculty, or another means.

Major advances in our understanding of faculty careers and labor markets could come with available periodic counts of faculty by department. The methods in Chapter 11, where we used data at three-year intervals, show that these counts would not have to be annual to be useful.

Appendix. Treatment of Departments in the Modeling

We have used institutional groupings, called "segments," throughout this book. Modeling every doctorate-granting university in a given field is impractical, yet the simulation requires a degree of institutional heterogeneity in order to deal with personnel shifts among institutions. This chapter describes the method we used to divide our universe of institutions into a separate set of segments for each of the thirteen science and engineering fields included in our study. Each field also contains three additional segments, which represent institutions that do not grant the doctorate. They are private liberal arts colleges, private comprehensive universities, and public colleges and comprehensive universities. The definitions for the non-doctoral segments are determined by the Carnegie classification system; hence the statistical procedures described here do not apply.

Variables

Data were obtained from NSF's CASPAR data system for 1980, the only year for which faculty are available by field and institution. In addition to the number of faculty (NFac), the following variables used in the segmentation analysis:

R&D: sponsored R&D dollars in thousands
Full-time science & engineering personnel

Expenditure on research equipment
Total graduate students in science & engineering

Full-time science & engineering personnel	Part-time graduate students in science & engineering
Science & engineering post-doctoral fellows	Masters degrees granted
Total doctoral degrees granted (NSF Survey)	Total doctoral degrees granted (NCES Survey)
Doctoral degrees granted to U. S. citizens	Bachelors degrees granted (whole institution)
Doctoral degrees granted to foreigners	Bachelors degrees granted (majors only)

The definitions are self-evident, except for the two surveys of doctoral degrees granted. One count comes from the "NSF Doctorate Survey File—Doctorate Institution Counts, Total Number of Doctorate Degrees." The other count comes from the file for "NCES Earned Degrees, Degrees Awarded.

Procedure

The segmentation procedure involves five steps:

1) Estimate missing values for *NFac* and *R&D*.
2) Perform factor analysis using the missing-data estimates where needed.
3) Calculate factor scores for the first factor.
4) Sort the institutions by their factor scores.
5) Allocate institutions into segments with roughly equal number of doctoral students. Steps 1–3 are performed without regard to type of institutional control. Steps 4 and 5 are performed separately for the public and public institutions.

We wish to establish each segment in a way such that student-faculty and similar "productivity" ratios will be relatively homogeneous within segments and heterogeneous among them. Therefore, we divide the variables by the number of faculty (*NFac*) before factor analysis. Thus, adjacent points on the real line defined by the first factor will represent institutions with what we might call "similar production processes." Without the transformation, the first factor would simply represent institutional size. By including a disproportionate number of variables related to graduate education and research, we increase the likelihood that the first factor will represent that set of activities. Indeed, one can view the factor analysis simply as a method for translating our battery of research and graduate education variables into a single interval-scaled dimension.

As might be expected, the more prestigious schools tend to

come out at the top of the private-sector table: for example, in the case of physics, Harvard, MIT, Cal. Tech, Cornell, and Stanford comprise segment 1. The array seems somewhat more mixed in the case of the public sector, with places like Oregon and SUNY, Albany ranking ahead of Berkeley. The predictive power in the case of public institutions is lower than for privates.

While we cannot separate the surprising from the anomalous, we are confident that, across the population of institutions, the factor analysis captures some meaningful relationships. The significance of the segmentation in resource allocation decisions summarized in Chapter 4, supports this conclusion.

Summary of segmentation by field

For reference, we include below in Table A.1 all thirteen fields of science and engineering included in this study. For each field we show its full name as listed in the NSF CASPAR data system, our four-letter abbreviation used in some tables and graphs, and a summary of the number of segments for private and public institutions. Tables A.2 and A.3 list each institution and give its segment position within public or private, counting from 1 as the most elite segment in both groups. Because of the small number of degrees in industrial engineering and limited data, we did not break institutions into segments, except for dividing private from public. Therefore, we do not list this field in the Tables A.2 and A.3. The other fields are shown.

Table A.1 Fields, Abbreviations, and Numbers of Segments

Field Name	Abbrev.	Number of Segments		
		Private	Public	Total
Biological Sciences	BIOL	3	7	10
Chemical Engineering	CENG	2	4	6
Chemistry	CHEM	4	6	10
Civil Engineering	CIVL	2	4	6
Computer Science	CSCI	2	2	4
Economics	ECON	4	6	10
Electrical Engineering	ELEC	4	6	10
Geological Sciences	GEOS	4	6	10
Industrial Engineering	INDU	1	1	2
Mathematics and Statistics	MATH	4	6	10
Mechanical Engineering	MECH	2	5	7
Physics and Astronomy	PHYS	4	6	10
Psychology	PSYC	4	6	10

Table A.2 Segment Positions: Private Universities

School	Biol-ogy	Chem. Eng.	Chem-istry	Civil Eng.	Comp. Sci.	Eco-nomics	Elec. Eng.	Geol. Sci.	Math. & Stat.	Mech. Eng.	Phys. & Astr.	Psychol-ogy
Adelphi University	3	2	3	2	2	4	4	3	3	2	4	1
American University	3	2	3	2	2	1	4	3	1	2	3	4
Andrews University	3	2	3	2	2	4	4	3	4	2	4	4
Baylor University	3	2	3	2	2	4	4	3	4	2	4	3
Biola University	3	2	3	2	2	4	4	3	4	2	4	3
Boston College	2	2	3	2	2	4	4	3	4	2	4	2
Boston University	3	2	3	2	2	4	4	3	2	2	4	3
Brandeis University	1	2	3	2	2	4	4	3	2	2	4	4
Brigham Young University	1	2	3	2	2	4	4	3	4	2	4	2
Brown University	3	2	3	2	2	3	4	2	2	2	3	4
California Institute of Technology	1	1	1	1	1	4	2	1	3	1	1	4
Carnegie Mellon University	3	2	3	2	2	4	3	3	4	2	4	4
Case Western Reserve University	2	2	1	2	2	4	4	3	4	2	4	3
Catholic University of America	1	2	3	2	2	3	3	3	4	1	3	2
Claremont Univ Center & Graduate School	3	2	3	2	2	4	4	3	4	2	4	4
Clark Atlanta University	1	2	3	2	2	4	4	3	4	2	4	4
Clark University	3	2	3	2	2	4	4	3	4	2	4	4
Clarkson University	3	2	3	2	2	4	4	3	4	2	4	4
Columbia University	2	1	1	1	2	2	3	1	4	1	2	2
Columbia University, Teachers College												
Cornell University	3	2	3	2	2	4	4	3	4	2	4	2
	2	2	1	2	1	1	3	3	3	2	2	3

University													
Dartmouth College	4	4	2	3	3	4	4	2	2	4	2	3	2	1
Drake University	4	4	2	4	3	4	4	2	2	4	2	3	2	3
Drexel University	4	4	1	4	3	4	4	2	2	4	2	3	2	2
Duke University	4	4	2	4	3	4	4	2	2	2	2	2	2	3
Duquesne University	3	4	2	4	3	4	4	2	2	4	2	3	2	2
Emory University	4	4	2	4	3	4	4	2	2	4	2	3	2	2
Florida Institute of Technology	1	4	2	4	3	4	4	2	2	4	2	3	2	3
Fordham University	2	4	2	4	3	4	3	2	2	4	2	3	2	3
George Washington University	2	4	2	1	3	4	3	2	2	3	1	3	1	1
Georgetown University	4	3	2	4	3	4	2	2	2	2	2	3	2	1
Hahnemann University	4	4	2	4	3	4	4	2	2	4	2	3	2	2
Harvard University	3	1	2	1	1	4	2	2	2	2	2	1	2	1
Hofstra University	1	4	2	4	3	4	4	2	2	4	2	3	2	3
Howard University	4	4	2	4	3	4	4	2	2	4	2	3	2	3
Illinois Institute of Technology	2	4	2	2	3	1	4	2	2	4	1	3	1	2
Johns Hopkins University	3	2	1	1	1	2	1	2	2	1	2	2	2	1
Lehigh University	4	4	2	4	3	4	4	2	2	4	2	3	2	2
Loma Linda University	4	4	2	4	3	4	4	2	2	4	2	3	2	2
Loyola University of Chicago	4	4	2	4	3	4	4	2	2	4	2	3	2	1
Marquette University	4	4	2	4	3	4	4	2	2	4	2	3	2	3
Massachusetts Institute of Technology	4	1	1	2	2	1	4	2	2	4	1	2	2	2
New School For Social Research	1	4	2	4	3	4	1	2	2	3	2	3	2	3
New York University	1	4	2	1	3	4	3	1	2	3	2	3	2	1
Northeastern University	4	4	2	4	3	4	4	2	2	4	2	2	2	2
Northwestern University	3	3	2	4	3	4	4	2	2	4	2	1	2	3
Nova University	2	4	2	4	3	4	4	2	2	4	2	3	2	2

Table A.2 Segment Positions: Private Universities (Continued)

School	Biol- ogy	Chem. Eng.	Chem- istry	Civil Eng.	Comp. Sci.	Eco- nomics	Elec. Eng.	Geol. Sci.	Math. & Stat.	Mech. Eng.	Phys. & Astr.	Psychol- ogy
Pepperdine University	3	2	3	2	2	4	4	3	4	2	4	2
Polytechnic University	3	1	3	1	1	4	2	3	4	2	4	4
Princeton University	2	1	1	2	2	4	1	3	1	2	3	4
Rensselaer Polytechnic Institute	3	2	3	1	2	4	4	3	3	2	4	4
Rice University	2	2	3	2	2	4	3	1	2	2	3	4
Rockefeller University	3	2	3	2	2	4	4	3	4	2	4	4
Saint John's University New York	2	2	3	2	2	4	4	3	4	2	4	2
Saint Louis University	2	2	3	2	2	4	4	1	2	2	4	3
Southern Methodist University	3	2	3	2	2	4	1	3	1	2	4	4
Stanford University	1	1	1	1	1	1	2	2	1	1	2	4
Stevens Institute of Technology	3	2	3	2	2	4	4	3	3	2	4	4
Syracuse University	2	2	3	2	2	3	4	3	4	2	3	3
Texas Christian University	3	2	3	2	2	4	4	3	4	2	4	4
Tufts University	2	2	3	2	2	3	4	3	4	2	4	4
Tulane University of Louisiana	3	2	3	2	2	4	4	3	4	2	4	4
Union Institute	2	2	3	2	2	4	4	3	4	2	4	1
University of Chicago	3	2	2	2	2	2	4	3	3	2	2	4
University of Denver	3	2	3	2	2	4	4	3	4	2	4	4
University of Miami	2	2	3	2	2	4	4	1	4	1	4	4
University of Notre Dame	2	1	2	2	2	1	3	3	4	2	4	3
University of Pennsylvania	3	1	2	2	2	3	3	3	4	2	4	4
University of Rochester	1	2	2	2	2	2	4	3	4	1	2	3
University of San Francisco	3	2	3	2	2	4	4	3	4	2	4	2

	Biol-ogy	Chem. Eng.	Chem-istry	Civil Eng.	Comp. Sci.	Eco-nomics	Elec. Eng.	Geol. Sci.	Math. & Stat.	Mech. Eng.	Phys. & Astr.	Psychol-ogy
University of Southern California	3	1	2	1	1	4	3	2	4	2	4	3
University of Tulsa	3	1	3	2	2	4	4	3	4	2	4	4
Vanderbilt University	2	2	3	2	2	2	4	3	3	2	4	2
Washington University	3	2	3	1	1	3	4	3	2	2	4	4
Yale University	3	2	2	2	2	3	4	3	3	2	3	4
Yeshiva University	2	2	3	2	2	4	4	3	4	2	4	1

Table A.3   Segment Positions: Public Universities

School	Biol-ogy	Chem. Eng.	Chem-istry	Civil Eng.	Comp. Sci.	Eco-nomics	Elec. Eng.	Geol. Sci.	Math. & Stat.	Mech. Eng.	Phys. & Astr.	Psychol-ogy
Arizona State University	2	4	6	4	2	6	4	7	5	5	5	4
Auburn University	7	1	7	4	2	6	2	7	6	5	6	6
Ball State University	7	4	7	4	2	6	6	7	3	5	6	5
Bowling Green State University	7	4	7	4	2	6	6	7	6	5	6	6
Clemson University	3	4	7	3	2	6	5	7	3	5	6	6
Cleveland State University	7	4	6	4	2	6	6	7	6	5	6	6
College of William and Mary	6	4	7	4	2	6	6	7	6	5	6	6
Colorado School of Mines	7	1	7	4	2	5	6	3	6	5	6	6
Colorado State University	6	4	5	3	2	6	5	5	6	5	5	6
CUNY Graduate School and University Center	7	4	7	4	1	4	6	7	6	5	6	6
East Texas State University	6	4	7	4	1	6	6	7	6	5	6	6
Florida Atlantic University	7	4	7	4	2	6	6	7	6	5	6	6
Florida State University	4	4	6	4	2	6	6	2	2	5	4	2

Table A.3 Segment Positions: Public Universities (Continued)

School	Biol-ogy	Chem. Eng.	Chem-istry	Civil Eng.	Comp. Sci.	Eco-nomics	Elec. Eng.	Geol. Sci.	Math. & Stat.	Mech. Eng.	Phys. & Astr.	Psychol-ogy
Georgia Institute of Technology	7	4	3	2	1	6	1	7	6	1	4	6
Georgia State University	7	4	7	4	2	6	6	7	6	5	6	3
Idaho State University	1	4	7	4	2	6	6	6	5	5	6	6
Illinois State University	7	4	7	4	2	6	6	7	6	5	6	6
Indiana State University, Terre Haute	6	4	7	4	2	6	6	7	6	5	6	5
Indiana University at Bloomington	3	4	3	4	2	3	6	7	2	5	5	3
Iowa State University	7	3	3	2	1	3	4	7	4	1	6	6
Kansas State University of Agri and App Sci	1	4	6	4	2	5	6	7	2	5	6	6
Kent State University	3	4	7	4	2	6	6	7	2	5	5	4
Louisiana State University and Agri & Mech Coll	7	4	7	4	2	6	6	7	6	5	6	6
Louisiana Tech University	7	4	7	4	2	6	6	7	6	5	6	6
Memphis State University	7	4	7	4	2	6	6	7	2	5	6	4
Miami University	6	4	7	4	2	6	6	7	5	5	6	6
Michigan State University	4	4	5	2	2	1	5	6	3	3	6	3
Middle Tennesse State University	7	4	7	4	2	6	6	7	6	5	6	6
Mississippi State University	4	4	7	4	2	6	6	7	6	4	6	6
Montana State University	1	4	7	4	2	6	6	7	6	5	5	6
New Mexico State University	7	4	7	4	2	6	5	7	6	4	4	6
North Carolina State University at Raleigh	5	1	6	4	2	6	6	6	6	4	6	4

Institution													
North Dakota State University	1	4	7	4	2	6	6	7	6	5	6	6	6
Northern Arizona University	7	4	7	4	2	6	6	7	6	5	6	6	6
Northern Illinois University	1	4	7	4	2	5	6	7	5	5	5	6	6
Ohio State University	4	4	2	1	1	2	5	7	3	4	3	3	1
Ohio University	3	4	7	4	2	6	4	7	5	5	5	6	4
Oklahoma State University	7	4	5	2	2	6	4	7	6	2	6	6	4
Old Dominion University	7	4	7	4	2	6	6	7	6	5	6	6	6
Oregon State University	4	4	5	4	2	5	3	1	5	4	5	6	6
Pennsylvania State University	5	1	1	1	2	4	5	5	3	4	3	1	2
Portland State University	7	4	7	4	2	6	6	7	6	5	6	6	6
Purdue University	3	2	2	2	1	1	2	6	3	3	3	4	5
Rutgers, the State University at New Brunswick	6	3	7	4	2	4	2	7	4	3	4	3	6
Southern Illinois University-Carbondale	6	4	7	4	2	6	6	7	6	5	6	3	3
SUNY at Albany	2	4	7	4	2	5	6	1	3	5	3	6	2
SUNY at Binghamton	6	4	6	4	2	6	6	4	2	5	2	6	5
SUNY at Buffalo	7	3	2	3	1	4	2	6	4	4	4	4	3
SUNY at Stony Brook	1	4	5	4	2	3	2	7	1	4	1	4	2
SUNY College of Envir Sci and Forestry	6	4	7	4	2	6	6	3	6	5	6	6	6
Temple University	2	4	7	4	2	6	6	7	6	5	6	6	1
Tennessee Technological University	7	4	7	4	2	6	6	7	6	5	6	6	6
Texas A & M University	2	2	4	2	2	1	4	3	1	5	1	6	4
Texas Tech University	1	4	3	3	2	6	6	6	6	5	6	6	3
Texas Woman's University	1	4	7	4	2	6	6	7	6	5	6	6	2

Table A.3 Segment Positions: Public Universities (Continued)

School	Biol-ogy	Chem. Eng.	Chem-istry	Civil Eng.	Comp. Sci.	Eco-nomics	Elec. Eng.	Geol. Sci.	Math. & Stat.	Mech. Eng.	Phys. & Astr.	Psychol-ogy
University of Akron	7	4	7	4	2	6	6	7	6	5	6	6
University of Alabama	7	4	7	4	2	6	6	7	3	4	6	5
University of Alabama at Birmingham	6	4	7	4	2	6	6	7	6	5	6	6
University of Arizona	1	4	5	2	1	5	5	3	6	3	1	4
University of Arkansas	7	4	7	4	2	4	6	7	6	5	6	6
University of California-Berkeley	1	3	1	1	2	1	1	2	1	1	2	4
University of California-Davis	6	4	6	4	2	6	6	5	5	4	5	5
University of California-Irvine	5	4	4	1	2	6	6	7	5	5	4	6
University of California-LosAngeles	5	4	3	4	2	2	6	4	5	5	1	3
University of California-Riverside	1	4	6	4	2	4	6	6	3	5	5	6
University of California-San Diego	5	4	3	4	2	6	6	2	5	5	3	6
University of California-San Francisco	7	4	6	4	2	6	6	7	6	5	6	6
University of California-Santa Barbara	4	4	4	4	2	5	1	5	2	4	4	5
University of California-Santa Cruz	3	4	6	4	2	6	6	5	2	5	1	6
University of Cincinnati	2	4	5	4	2	2	2	7	6	3	5	6
University of Colorado at Boulder	2	4	7	4	1	3	6	4	3	5	2	2
University of Connecticut	1	4	7	4	2	4	4	7	5	3	5	3

University												
University of Delaware	1	2	4	4	2	6	4	2	6	4	6	6
University of Florida	6	4	4	1	2	4	5	6	5	3	5	5
University of Georgia	3	4	3	4	2	5	6	6	2	5	6	1
University of Hawaii at Manoa	7	4	6	4	2	4	6	6	6	5	6	3
University of Houston-University Park	2	1	5	3	2	6	3	6	2	3	6	5
University of Idaho	1	4	6	4	2	6	6	7	6	4	6	6
University of Illinois at Chicago	1	4	3	4	2	6	6	7	4	5	6	6
University of Illinois at Urbana-Champaign	4	2	1	3	1	3	3	2	4	2	2	3
University of Iowa	4	2	5	4	2	6	5	7	4	4	4	6
University of Kansas	3	4	5	4	2	5	5	7	6	5	6	5
University of Kentucky	7	4	7	4	2	3	6	6	3	4	6	1
University of Louisville	7	4	7	4	2	6	6	7	6	4	5	5
University of Maine at Orono	3	4	7	4	2	6	6	7	6	5	6	6
University of Maryland at College Park	4	4	6	4	2	4	5	7	6	5	3	4
University of Maryland Baltimore County	7	4	7	4	2	6	6	7	6	5	6	6
University of Massachusetts at Amherst	2	4	2	3	2	5	6	7	6	5	5	6
University of Michigan at Ann Arbor	5	4	4	2	1	5	4	3	4	5	5	5
University of Minnesota at Twin Cities	3	3	4	2	2	2	4	7	5	5	5	1
University of Mississippi	1	4	7	4	2	6	6	7	6	1	6	1
University of Missouri, Columbia	4	4	7	1	2	2	1	4	6	3	6	2

Table A.3 Segment Positions: Public Universities (Continued)

School	Biol- ogy	Chem. Eng.	Chem- istry	Civil Eng.	Comp. Sci.	Eco- nomics	Elec. Eng.	Geol. Sci.	Math. & Stat.	Mech. Eng.	Phys. & Astr.	Psychol- ogy
University of Missouri, Kansas City	6	4	6	4	2	6	6	7	6	5	6	6
University of Missouri, Rolla	7	4	7	4	2	6	5	7	6	4	6	6
University of Missouri, Saint Louis	7	4	7	4	2	6	6	7	6	5	6	6
University of Montana	1	4	7	4	2	6	6	7	5	5	6	5
University of Nebraska at Lincoln	6	4	4	4	2	3	6	7	5	5	6	2
University of Nevada-Reno	7	4	6	4	2	6	6	6	6	5	6	5
University of New Hampshire	3	4	5	4	2	6	6	7	5	5	6	6
University of New Mexico	5	4	7	1	2	6	5	3	6	4	6	5
University of New Orleans	7	4	7	4	2	6	6	7	6	5	6	6
University of North Carolina at Chapel Hill	7	4	4	4	2	5	6	6	5	5	5	3
University of North Carolina at Greensboro	7	4	7	4	2	6	6	7	6	5	6	5
University of North Dakota	6	4	7	4	2	6	6	7	6	5	6	4
University of North Texas	1	4	6	4	2	6	6	7	6	5	4	2
University of Northern Colorado	7	4	7	4	2	6	6	7	2	5	6	6
University of Oklahoma, Norman Campus	7	1	7	1	2	6	2	2	6	5	6	4
University of Oregon	7	4	5	4	2	6	6	7	4	5	1	4
University of Pittsburgh	5	3	2	2	1	4	3	6	5	3	5	4
University of Rhode Island	6	4	6	4	2	6	6	4	6	5	6	5

University of South Carolina at Columbia	7	4	4	4	2	6	6	5	4	5	5	3
University of South Dakota	7	4	7	4	2	6	6	7	6	5	6	4
University of South Florida	1	4	7	4	2	6	6	7	6	5	6	5
University of Southern Mississippi	1	4	7	4	2	6	6	7	6	5	6	1
University of Tennessee at Knoxville	3	2	6	2	2	6	6	6	2	5	6	1
University of Texas at Arlington	7	3	7	3	2	3	3	7	2	5	6	6
University of Texas at Austin	5	1	2	3	2	4	4	6	3	4	3	1
University of Texas at Dallas	5	4	7	4	2	6	6	5	2	5	6	6
University of Toledo	7	4	7	4	2	6	6	7	6	4	6	5
University of Utah	3	2	3	4	2	4	6	5	6	5	4	2
University of Vermont	1	4	6	4	2	5	5	7	6	5	6	3
University of Virginia	3	4	5	4	2	4	6	7	2	5	4	6
University of Washington	2	4	5	4	1	6	5	1	5	5	4	6
University of Wisconsin-Madison	2	3	1	4	1	5	2	1	1	2	3	1
University of Wisconsin-Milwaukee	6	4	6	4	2	6	6	7	5	5	6	1
University of Wyoming	1	4	7	4	2	6	5	5	1	5	6	5
Utah State University	5	4	7	4	2	5	6	7	6	5	6	6
Virginia Commonwealth University	2	4	7	4	2	6	6	7	1	5	6	4
Virginia Polytechnic Institute and State Univ	6	4	6	4	2	3	5	7	5	2	6	6
Washington State University	3	4	7	4	2	4	6	4	6	5	4	2
Wayne State University	3	4	5	4	2	6	6	7	6	4	6	2
West Virginia University	4	3	6	4	2	6	6	6	6	4	6	5
Western Michigan University	5	4	7	4	2	6	6	7	5	5	6	1

Bibliography

Amemiya, T. (1985). *Advanced econometrics*, Cambridge, MA: Harvard University Press.

Atkinson, R. C. (1990, April 27). Supply and demand for scientists and engineers: A national crisis in the making, *Science, 248*, 425–432.

Beightler, C. S., Phillips, D. T., & Wilde, D. J. (1979). *Foundations of optimization*. Englewood Cliffs, NJ: Prentice-Hall.

Bowen, H. R. (1980). *The costs of higher education: How much do colleges and universities spend per student and how much should they spend?* San Francisco, CA: Jossey-Bass.

Bowen, H. R., & Schuster, J. H. (1986). *American professors: A national resource imperiled.* New York, NY: Oxford University Press.

Bowen, W. G., & Sosa, J. A. (1989). *Prospects for faculty in the arts and sciences.* Princeton, NJ: Princeton University Press.

Bowen, W. G., Lord, G., & Sosa, J. A. (1991, February). Measuring time to the doctorate: Reinterpretation of the evidence. *Proceedings of the National Academy of Sciences, 88*, 713–717.

Bowen, W. G., & Rudenstine, N. L. (1992). *In pursuit of the PhD.* Princeton, NJ: Princeton University Press.

Carroll, S. (1995). *Projecting California's fiscal future*, MR-570–LE, Santa Monica, CA: RAND.

College and University Personnel Association. (1982, June 2). Faculty salary survey. *The Chronicle of Higher Education.*

College and University Personnel Association. (1987, April 29). Faculty salary survey. *The Chronicle of Higher Education.*

College and University Personnel Association. (1993, March 31). Faculty salary survey. *The Chronicle of Higher Education.*

Committee on Science, Engineering, and Public Policy. (COSEPUP) (1995). *Reshaping the graduate education of scientists and engineers.* Washington, DC: National Academy Press.

Hillier, F. S., & Lieberman, G. J. (1990). *Introduction to operations research* (fifth ed.). New York, NY: McGraw-Hill Publishing Company.

Hopkins, D. S. P., & Massy, W. F. (1981). *Planning models for colleges and universities.* Stanford, CA: Stanford University Press.

James, E. (1982). How nonprofits grow: A model. *Journal of Policy Analysis, 2,* 350–366.

James, E. (1990). Decision processes and priorities in higher education. In S. A. Hoenack & E. L. Collins (Eds.), *The economics of American universities,* 77–106. Albany, NY: SUNY Press.

Lehmann, E. L. (1991). *Theory of point estimation.* Pacific Grove, CA: Wadsworth & Brooks/Cole Advanced Books and Software.

Massy, W. F., & Wilger, A. (1995, July-August). Faculty productivity. *Change.*

Massy, W. F., Wilger, A., & Colbeck, C. (1994, January-February). Overcoming 'hollowed collegiality'. *Change,*10–20.

Massy, W. F., Montgomery, D. M., & Morrison, D. G. (1974). *Stochastic models of buying behavior.* Cambridge, MA: MIT Press.

McGuire, M. D., & Price, J. A. (1989, May 29). *Faculty replacement needs for the next 15 years: A simulated attrition model.* Paper presented at the Annual Forum of the Association for Institutional Research, Baltimore, MD.

National Research Council. (1965–1991). *Survey of earned doctorates: 1965–1991.* Washington, DC: National Research Council.

National Science Foundation. (1993). *Science and engineering doctorates: 1960–91.* NSF 93–301. Detailed Statistical Tables. Washington, DC, 153–154.

National Science Foundation (CASPAR) *Computer Aided Science Policy Analysis and Research Database System*, developed for the National Science Foundation by Quantum Research, Bethesda, MD, various years.

National Science Foundation. *Science and Engineering Statistical Data System* (SESTAT), various years.

Rao, C. R. (1973). *Linear statistical inference and its applications.* New York, NY: John Wiley.

Romer, P. M. (1990, supplement). Endogenous technical change. *Journal of Political Economy, 96.*

Russell, S. H., et al. (1990). *1988 National Survey of Postsecondary Faculty (NSOPF-88): A Descriptive Report of Academic Departments in Higher Education Institutions*, NCES 90–339, US Department of Education, Office of Educational Research and Improvement, National Center for Education Statistics.

Russell, S. H., et al. (1991). *1988 National Survey of Postsecondary Faculty (NSOPF-88): Profiles of Faculty in Higher Education Institutions*, NCES 91–389, US Department of Education, Office of Educational Research and Improvement, National Center for Education Statistics.

Schwefel, H. P. (1977). *Numerical optimization of computer models.* New York, NY: John Wiley & Sons.

Stata Corporation. (1993). *Stata reference manual: Release 3.1* (6th ed.). College Station, TX: Stata Corporation.

Stata Corporation. (1993). *Stata (Release 3.1)* [Computer software]. College Station, TX: Stata Corporation.

Tuckman, H., Coyle, S., & Bae, Y. (1990). *On time to the doctorate: A study of the increased time to complete doctorates in science and engineering.* Washington, DC: National Academy Press.

Western Interstate Commission for Higher Education. (WICHE). (1992, March). *Bringing into Focus the Factors Affecting Faculty Supply and Demand: A Primer for Higher Education and State Policymakers.* Boulder, CO: WICHE.

Wolfram, S. (1991). *Mathematica.* Redwood City, CA: Addison-Wesley.

Index